FIVE RINGS
ILLUSTRATED

amber
BOOKS

"Never stray from the Way"

– Miyamoto Musashi

FIVE RINGS
ILLUSTRATED

THE CLASSIC TEXT ON MASTERY IN SWORDSMANSHIP,
LEADERSHIP AND CONFLICT: A NEW TRANSLATION

Miyamoto Musashi

Translated by Maisy Hatchard

Reprinted in 2023, 2024

Amber Books Ltd
United House
North Road
London N7 9DP
United Kingdom
www.amberbooks.co.uk
Instagram: amberbooksltd
Facebook: amberbooks
Twitter: @amberbooks
Pinterest: amberbooksltd

Copyright © 2022 Amber Books Ltd

All rights reserved. No part of this work may be reproduced, stored in a retrieval system, or transmitted in any form or by any means, electronic, mechanical, photocopying, recording, or otherwise, without the prior permission of the copyright holder.

ISBN: 978-1-83886-217-6

Translated by Maisy Hatchard

Project Editors: Sarah Uttridge & Michael Spilling
Design: Rick Fawcett & Keren Harragan
Picture Research: Terry Forshaw

Printed and bound in China

TRADITIONAL CHINESE BOOKBINDING

This book has been produced using traditional Chinese bookbinding techniques, using a method that was developed during the Ming Dynasty (1368–1644) and remained in use until the adoption of Western binding techniques in the early 1900s. In traditional Chinese binding, single sheets of paper are printed on one side only, and each sheet is folded in half, with the printed pages on the outside. The book block is then sandwiched between two boards and sewn together through punched holes close to the cut edges of the folded sheets.

Contents

Introduction 6

地の巻 **The Scroll of Earth** 8

水の巻 **The Scroll of Water** 44

火の巻 **The Scroll of Fire** 84

風の巻 **The Scroll of Wind** 128

空の巻 **The Scroll of Expanse** 154

Glossary 157

Index 159

Introduction

The book in your hands is the culmination of a life's training – a gifted samurai's teachings on how practice and perspective are essential in any path that we follow, not only in physical combat, but in our everyday lives and chosen skills. Miyamoto Musashi's *Five Rings* differs from other works and teachings of its time in several ways. He intentionally kept his writing devoid of religion, he incorporates and gives due importance to other occupations, and he was concerned with ensuring that all his pupils, no matter their means, were able to follow his teachings.

This new translation follows his desire for accessibility, and is therefore written with context and readability in mind. Whether you picked up this book because you are interested in Japanese history, in samurai culture, in the sensational aspects of Japan, or because you liked the cover, I hope to take you on a journey with Musashi's philosophy. *Five Rings* is translated in a way that retains accuracy while adding context where it's needed to avoid distracting the reader with footnotes. I hope that you find this an enjoyable balance between an historical work and a profound yet satisfying read.

Musashi himself has gained legendary status over time. Born in 1582 in modern day Hyōgo prefecture in the southwest of Japan's main island, he lived during a pivotal point in samurai life. We often romanticize the Edo period of 1603–1868 as a peaceful golden age of wandering samurai, art creation and economic growth. However,

Musashi also witnessed this period's slow stagnation of the samurai class, even noting in his text that those carrying the swords might not even know their uses.

The Five Rings, the elements of Earth, Water, Fire, Wind and Expanse that he splits the work into, are more than just his record of his school's combat teachings. It is an attempt to help those that are lost in their training to find a new perspective, to give guidance in applying strategy across a broad range of vocations, and to keep what he saw as a dying samurai mindset alive. Since his death in 1645 Musashi's legacy has continued to achieve just that. On top of the schools of martial arts that still exist today because of him, Musashi has inspired countless films and stories, as well as multitudes of games.

The reason his work continues to influence us some 375 years later, is because of how widely applicable it is. Anyone can learn from Musashi's teachings. The mindset that he encourages his students to adopt, and his strategies for coming out on top, apply not only to their original audience of martial arts students and military leaders, but also to corporate heads, athletes and even casual readers. Musashi teaches a full spectrum; from the physical details of sword technique itself, to thinking in a way that will bring efficiency into anything that you do.

So whatever your reason for picking up this book today, I think you'll enjoy a glimpse into what being a top-tier samurai meant to Musashi and that you'll find it useful, if not for the sword techniques, then for the strategy he weaves throughout.

地の巻

The Scroll of Earth

Introduction

I believe it is finally time to put my Strategic Way of the Warrior, or *Niten-Ichiryū*, as I have called it, into writing for the first time. It is the beginning of October, in the year *Kan'ei* 20 (1643). I have climbed Mt. Iwato in Kumamoto prefecture in Kyushu. I have prayed to heaven, given worship to Kannon, and I now sit here facing Buddha. I am a warrior native to Harima, by the name of Shinmen Musashi-no-Kami Fujiwara-no-Genshin, and I have 60 years behind me.

I have devoted myself to the Way of the Warrior since I was young, following my first real duel at the age of 13. My opponent was a *Shintō-Ryū* user named Arima Kihei. At 16, I bested Akiyama of Tajima, a warrior known for his strength. At 21, I headed for Kyoto, where I had encounters with many of the finest warriors under the sun, never once failing to claim victory. I continued to travel to various different places and met warriors of many schools, but I remained undefeated in more than 60 battles. That is how I spent my younger years, from age 13 to age 28 or 29.

In a scene set in snowy hills, the noted Osaka actor Arashi Rikan II (1788–1837) plays a role inspired by the famous warrior and swordsman Miyamoto Musashi.

When I passed 30 years of age, I looked back over my past contests only to realize that my victories were not due to a mastery of the Strategic Way of the Warrior. They were more likely to have been due to an innate talent for the Way that stopped me from straying too far from divine truths, or because of some flaw in the opponents from the other schools. Since that realization I have worked tirelessly, morning and night, to forge an understanding of deeper truths.

It was not until I reached 50 years of age that I finally came to truly understand the Way of the Warrior. Since reaching a perfect understanding, I have not needed to spend my years pursuing this path. Instead, the truths revealed to me by the Way of the Warrior spur me along the paths of other disciplines. In all of these I have no master.

Now, as I write this book, I do not borrow the ancient teachings of Buddhism or Confucianism, nor do I reference historical war chronicles or military strategies. This is simply a revelation of the true spirit of *Niten-Ichiryū*. I take up my brush-pen at dawn on the tenth day of the tenth month, with Kannon and Heaven as my witness, and begin my chronicle.

The Strategic Way

The Strategic Way is, in essence, the samurai's charter. It is of utmost importance for generals to practise this charter, and for their soldiers to know it. Yet there is currently not a single samurai on earth who truly holds an understanding of the Way of the Warrior.

兵法の本来のありよう

Of all the Ways in the world, each and every practised path is followed as its practitioners please. First and foremost is the Buddhist path to salvation, and then the way of words as paved by Confucianism. The path of doctors and their ways to heal all manner of illnesses, or poets teaching their ways of *waka* (poetry). Tea ceremony practitioners, archers, and the myriad other artistic disciplines: all are followed because someone desires to do so.

However, there are few who desire to follow the Way of the Warrior. The true path for the Way of the Warrior is a wholehearted dedication to both the literary and military ways, which are known collectively to the warrior as *Bunbu-Ryōdō*. Even if a samurai is ill-suited to this path, it is their duty to strive for the Way of the Warrior as much as their social standing allows.

Generally speaking, however, I gather that the current generations of samurai believe that their path is simply a wholehearted acceptance of death. But accepting death is not something unique to the warrior class. Priests, women, farmers, and even the lowest class of society know their obligations and contemplate shame. Their resolution in the face of death makes them not so different from warriors.

For samurai who practise the Way of the Warrior, the most fundamental basis of the path involves exceeding others in anything they do. Whether it is success in crossing swords with a single opponent, or winning a battle against many, exceeding others is how we gain prestige, and how

This 18th century print shows the actor Nakamura Shikan as a samurai about to string an arrow amidst a snowy landscape.

In this 18th century print, actor Ichikawa Danzō III plays the role of a samurai about to draw his long two-handed katana *sword.*

we make a name for ourselves and our lords. All of this can be found in the excellence of the Way of the Warrior.

There are likely those who can't see how studying the Way of the Warrior would be useful in a real battle. To them, I emphasize that the true Way of the Warrior teaches precisely that your craft should be honed so that it can be used at any time and in any circumstance.

1. What 'Strategic Way of the Warrior' Means

Whether in China or Japan, those who understand and follow the path have come to be called masters of the Strategic Way. As a samurai, it is imperative to study this path.

However, in recent years, the so-called 'strategic masters' making their way in the world have actually mastered nothing more than the sword. In fact, it wasn't long ago that attendants of the Hitachi province shrines, Kashima and Katori, created schools that they claimed to be the teachings of deities, preaching their swordwork far and wide.

In contrast, the Strategic Way is a practical and beneficial Art, of the Ten Skills and Seven Arts of old. It is much more than just swordwork, and to think of it in such a limited way eliminates all its practicality. True comprehension of strategic swordwork will not come from mere sword skills alone. Nor, of course, will the ability to follow the principles of combat.

On the whole, the people of the world turn their various crafts into wares for sale, themselves and their tools included. Of the two, the fruit and the flower, the world is filled with more flowers than actual fruit. This is especially true for the Way of the Warrior. There is an obsession with colourful displays, by people showing off skills as though they are forcing flowers into bloom. People promote this and that dojo but have no combat experience. They teach one path, or learn another, to try to reap the rewards of victory in battle. All these trends show the truth behind the well-known saying: 'unrefined strategy yields nothing but harm'.

Broadly speaking, there are four Ways to make a living in the world: that of the farmer, merchant, samurai or artisan.

The first Way is the path of agriculture. Farmers ready their tools to spend the year working with the seasons. They keep a close eye on the changes in spring, summer, autumn and winter. This is the Way of the farmer.

The second Way is the path of commerce. *Sake* brewers craft their wines and make their way in the world from the profits, contingent on the good or bad quality of their product. The path of commerce is entirely one of making a profit to earn a living. This is the Way of the merchant.

The third Way is the path of the Warrior. It is referred to as the warrior's path because a samurai not only makes their own equipment, but also understands the usefulness of each piece. Samurai who do not care to fashion their

This late-19th century engraved illustration shows a Japanese farmer dressed in winter clothing made from straw.

own equipment and therefore don't understand the merits of its use are lacking the right attitude.

The fourth Way is the path of the artisan. In the Way of carpentry, carpenters make their way in the world by skilful use of various tools; building pieces exactly to plan, according to their precise rules. They are constantly honing their skills.

These are the four paths of the samurai, farmer, artisan and merchant.

We shall take carpentry as an example for the Strategic Way. It is an easy comparison to draw since both are related to 'houses'. There is the imperial house, the warrior houses, and the Four Noble Houses of great lineage. We can mark the dissolution and endurance of houses. Schools and traditions are referred to as houses. It is because the label 'house' is used in both the carpenter's and the samurai's path that I make the comparison.

The word 'carpenter' itself is written with the characters for 'great' and 'artisan craft', and though the Strategic Way is written differently, it is a great artisanal craft in itself. To put it another way, both require prudence and methodical planning, which is why I draw the comparison with the path of carpentry.

Think carefully about what is written in these scrolls if you intend to learn the Strategic Way of the Warrior. Practise industriously, with a teacher as the needle to the student's thread.

2. *The Strategic Way of the Warrior as Equated to Carpentry*

A general-in-chief discerns between right and wrong, upholds the laws of the country and, in the same way as does a chief carpenter, knows the ways of their lord's house. This is the path to mastery. A chief carpenter remembers the dimensions of temple buildings, knows the plans for palaces, and manages subordinates to build each one. In this way, mastery of carpentry and mastery of combat are the same concept.

When building a house, a carpenter must decide which wood to use and where. Beautiful, straight timber with no knots will become the exterior supports. Slightly knotted but straight and strong timber will be used as the rear support pillars. Unknotted but slightly weak timber with a beautiful grain will be used for decorative beams or sliding door rails and frames, or the doors themselves. As long as the house is carefully assessed for which strength of wood is needed in each part, even knotty, twisted timber can be useful for the construction of a long-standing house. If there are pieces in the timber that are simply too knotty, twisted and weak for the house, they can still be used as scaffolding or firewood.

When a chief employs carpenters, they must know the level of competence of each employee and be able to place them accordingly. One person for the floorboards, one for the doors and frames, one for the door thresholds, another for lintels, another for the ceiling. Even the less skilled workers can be made to lay the cross-pieces beneath the floorboards, and the truly unskilled can be set to

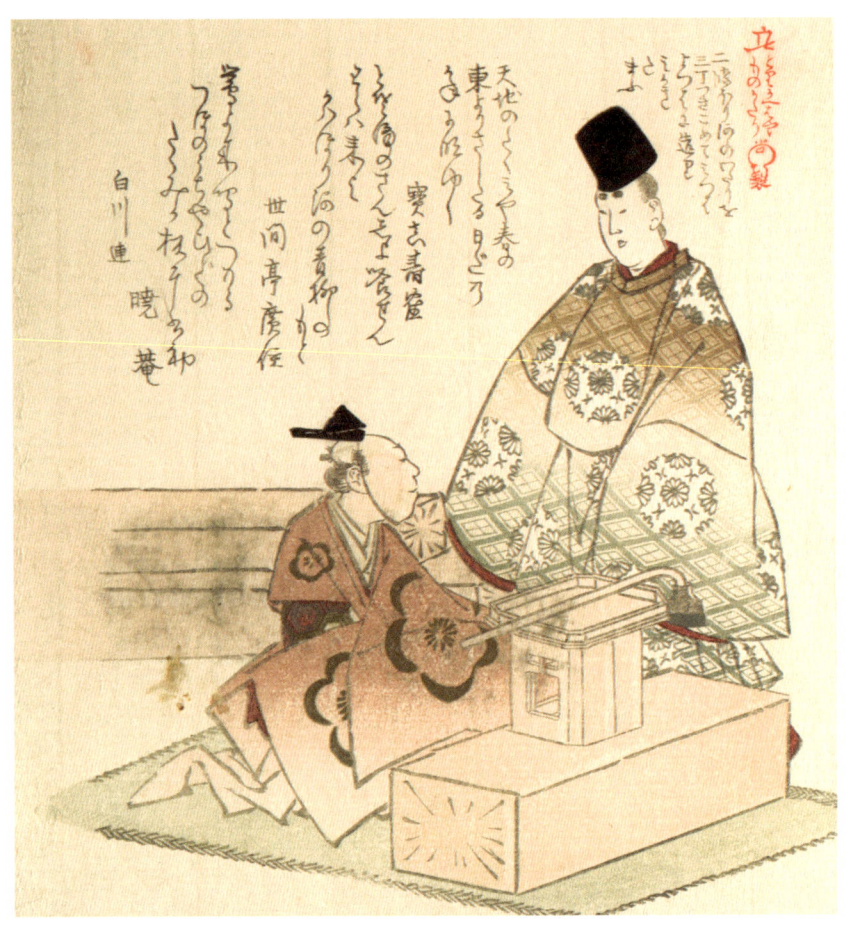

A young nobleman is tended to by a carpenter in this 19th-century woodblock print.

sanding. The job will go smoothly when everyone is given appropriate tasks.

Principles such as moving efficiently to make progress, cracking down on laziness, perceiving the priorities and motivations of your people, being a catalyst for energy in the work, and understanding limitations are all things that should be at the front of a chief's mind. The principles of the Strategic Way are one and the same.

3. The Strategic Way of the Warrior

Soldiers are quite a lot like trainee carpenters. Sharpening one's own tools, making various replacement tools and carrying them on their person in their toolboxes is par for the course for a trainee carpenter. They follow instructions from the chief carpenter to chop the wood for pillars and beams, plane the shelves and floorboards, and carve intricate ornamentation. They dutifully make sure that everything, from the smallest of intricate details to the long passageways through which horses are ridden, is perfectly finished to plan. The carpenter who commits the tricks of the trade to muscle memory and has a strong handle on construction is sure to become a master builder.

It is important for carpenters to keep well-sharpened tools. Any spare time should be used fundamentally to maintain these tools. A carpenter uses those tools to make everything from shrine cabinets to bookshelves, desks, lanterns, chopping boards and even pot lids, so maintenance is of utmost importance. It is the same

thing for soldiers. This is something that should be well thought over.

A carpenter must take the utmost care to make sure that the wood doesn't warp, that the joints are aligned perfectly and that everything is beautifully sanded so it doesn't need polishing up later. Those who intend to study the Strategic Way must be similarly meticulous. Commit everything here to memory, scrutinizing the content word by word.

4. *The Composition of This Book of Strategy and its Five Scrolls*

This book is split into five Ways. The main elements of each Way are revealed in the scrolls of Earth, Water, Fire, Air and Expanse.

In the Scroll of Earth, I talk about the overarching principles of the Strategic Way from the standpoint of my school. One cannot fathom the true meaning of the path with sword skills alone. One must also perceive small details from among the big, and arrive at the deeper places from the shallow. To build stable ground upon which your path may run straight and true, I give the first scroll the name of 'Earth'.

Second is the Scroll of Water. The heart should be as the water flows. Square or circular, water fits its container; drop or ocean, water grows and shrinks. I will write out the facets of *Ichiryū* so that they are as clear as the clear blue colour of water. If you can discern the many truths

Master swordsmith Goro Masamune forges a katana *with help from an assistant. The clothing is an indicator of the ritualistic practice of crafting swords.*

一事によって万事を知ること

of swordsmanship and master the ability to freely triumph over one person, you will be able to triumph over anyone in the world. The frame of mind to win victory over someone is the same whether your opponents number one thousand or ten thousand. A general of the army applying small concepts on the large scale is the same as making a great Buddha statue with a tiny scale model.

I cannot describe it fully in writing, but the principle of winning a battle in accordance with the Strategic Way is this: appreciate the ten thousand things that are revealed from just one. This is the heart of my school, and I record it in this Scroll of Water.

Third is the Scroll of Fire. I write about the combat forms in this scroll. Whether a fire is an inferno or a small flame, it always has a blazing spirit. That is why I use it to write about battle. The way of war is the same for a one-on-one fight as it is for a ten thousand-a-side battle. Some situations require us to think big and others to think small. Which is most fitting requires careful consideration.

The bigger picture is easy to see, but the smaller details are not. By this I mean that a large group will not change its goal too quickly, so it is easy to keep an eye on the big picture, but a single person may do as their individual thinking dictates, changing course at rapid speed and making it hard to predict the smaller things. Always keep this in mind.

The matters in the Scroll of Fire describe mere moments in battle, but an important part of the Way of the Warrior

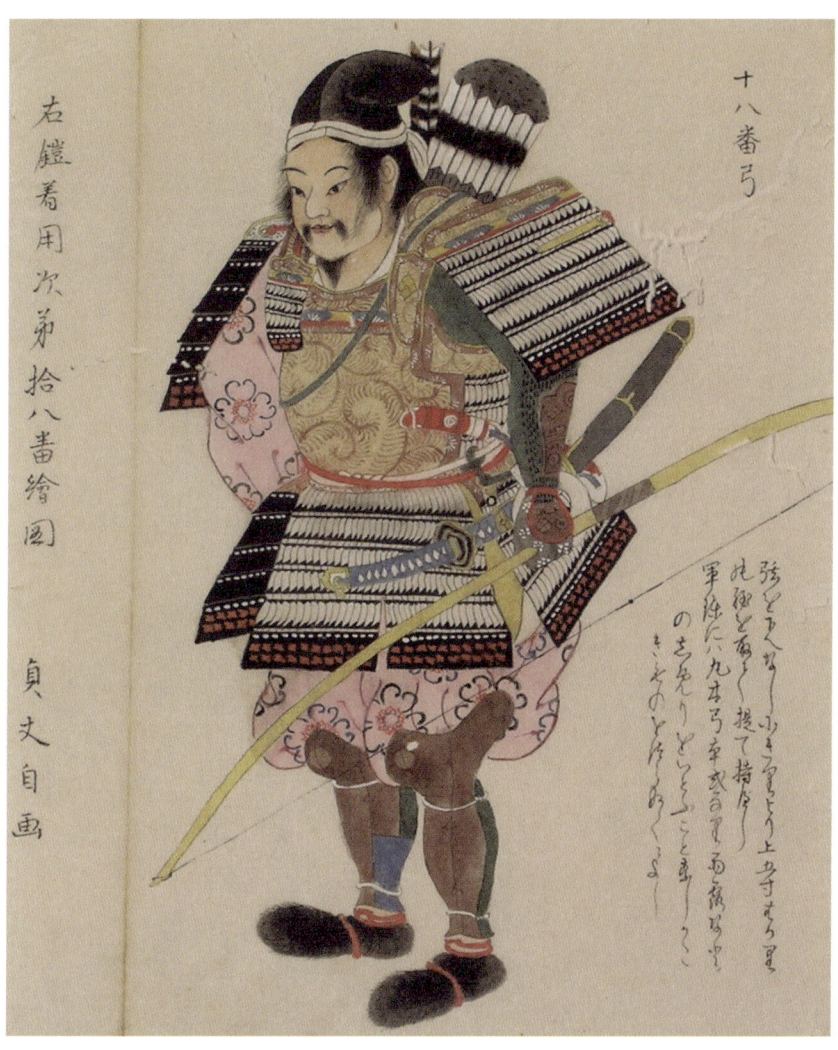

Part of a series of 23 paintings illustrating the steps in dressing a samurai warrior in Yoroi armour (Japan, 1780).

Actor Matsumoto Kōshirō IV plays a samurai warrior, here armed with two swords – the longer katana *and the shorter* wakizashi.

is becoming accustomed to keeping an unwavering spirit, day in, day out. This is why I write about the contest in battle in the Scroll of Fire.

Fourth is the Scroll of Wind. In this scroll I write not about my own *Niten-Ichiryū*, but about the various schools of the world. The 'winds of change' bring about current trends, blow away the old, and howl through the houses of tradition. Thus the Scroll of Wind is a reflection of the various strategies in the world, and the way their schools tend to do things. You cannot truly understand yourself without knowledge of the world around you.

In following the various Ways and arts, one may take a wayward tangent. Even if you believe you will stay on the correct path by remaining diligent, if your mind wanders, then so shall you. Perspective down the line will reveal that your path is not the correct one. On your journey along the path, even a slight meander partway will lead to a huge digression. No matter what it is, too much of something is as bad as too little, and this requires careful thought.

The worldwide assumption that other schools are only geared towards sword skill is absolutely correct. The principles of victory in battle and the techniques practised in my *Niten-Ichiryū* are vastly different from other schools. In order to give perspective on the other paths and trends, I write this Scroll of Wind.

Fifth is the Scroll of Expanse. I write of the expanse, but how can I define its depths or threshold when it

speaks of absence? Once you learn the truth of the Way, you are unbound by it. The Way of the Warrior itself is instinct and freedom, and it brings with it an unconscious and mysterious power to understand the rhythm of the moment. Your strikes will come instinctively of their own accord, and hit perfectly. This is the essence of the Expanse. I put down into words how one can naturally fall upon the true Way in the Scroll of Expanse.

5. Calling my Niten-Ichiryū School Nitō-Ryū (Two Swords as One)

I chose to name my school *Nitō* because wearing two swords on one's belt is the duty of warriors; generals and soldiers alike. In the past, the longer sword was called a *tachi*, and the shorter, a *katana*. Now they are called *katana* and *wakizashi*, the latter being the shortest sword. Anyone of samurai status wears these two swords, though the reason for doing so is not documented in great detail. In our nation wearing two swords at the waist is a symbol of the Way of the samurai, whether the one wearing them knows their uses or not. To make known to the world the advantages of using both a *tachi* and a *wakizashi*, I propose the name of *Nitō-Ryū*.

Other weapons, like the spear, and pole weapons, such as the *naginata,* are additional, spare arms. The proper way to begin practising the *Niten-Ichiryū* Way of the Warrior is to start training with a *tachi* and a *wakizashi* in each hand. This is intrinsic to my teaching.

When you are about to give up your life in battle, your swords should not be left in your belt as a last resort.

This centre panel of a triptych by Enryosai Shigemitsu shows an actor playing the role of expert swordsman Miyamoto Musashi.

A carved ivory tachi *ceremonial sword, circa 1900.*

You want to make full use of them. Dying with your weapons uselessly dangling from your belt is unacceptable.

However, having a sword in each hand does make it difficult to manipulate them freely. *Nitō-Ryū* is about becoming comfortable with the *tachi* as a one-handed weapon. Larger weapons like the spear and *naginata* are customarily two-handed, but both the longer *tachi* and the *wakizashi* are weapons that should be held one-handed.

Wielding a *tachi* with both hands is disadvantageous when on horseback or when running, whether on waterlogged land, on rocky footing or difficult roads, or in a fray of people. Wielding a *tachi* in both hands is not correct form, given that holding a spear, bow or other weapon in the left hand would mean that you could only wield the *tachi* one-handed. If it is really too difficult to cut someone down with one hand, then you may resort to two. It is not a difficult concept to grasp.

Start by wielding two swords simultaneously in the *Nitō* way, and work to gain the muscle memory for a one-handed *tachi* swing. Anyone taking up the *tachi* in one hand will find it heavy and cumbersome at first. Any weapon is the same; a bow is hard to draw and a *naginata* is hard to swing. But once you get to know the bow well your draw becomes stronger. Similarly, once you get used to swinging a *tachi*, you will understand its path and it will become easier to swing.

However, the path of the *tachi* is not simply to swing it faster. You can study this in the Scroll of Water, but the

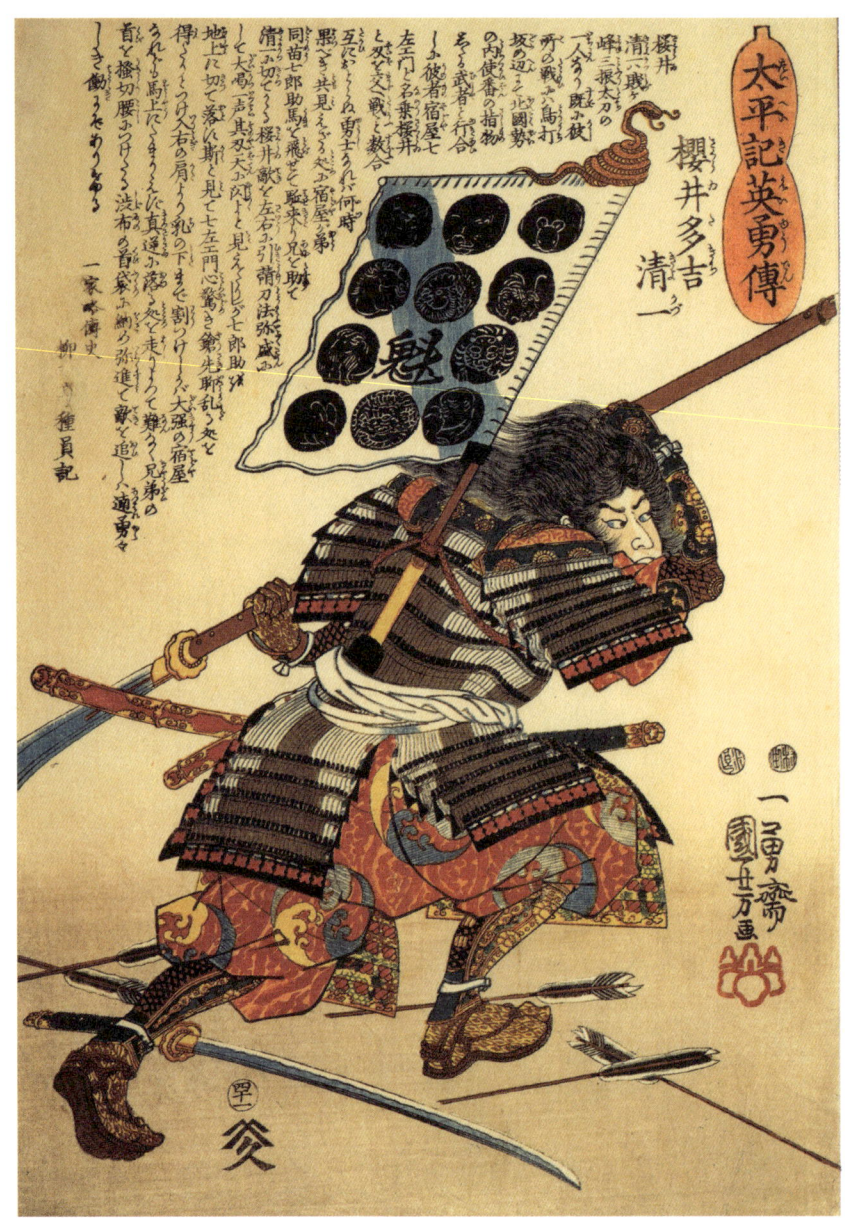

Sakuri Takichi Kiyokazu striking a blow with his naginata, *by artist Utagawa Kuniyoshi (1797–1861).*

fundamental idea is to use the right sword for the right situation; the *tachi* for open areas and the *wakizashi* for confined ones. In *Niten-Ichiryū*, we undertake to win the fight regardless of whether the sword is long or short, so I do not prescribe a length for the swords we use. Pushing oneself to gain the victory with any length sword is the Way of my *Niten-Ichiryū* school.

There are situations in which it is better to wield two swords instead of one, such as when you are one against many, or against a foe who is backed into a corner. I won't write this out in detail here, but you must appreciate the ten thousand things that are revealed from just one. Once you realize the principles of victory you will not miss a single thing. This is something you ought to reflect on thoroughly.

6. Understanding the Principles Behind the Two Words in a 'Strategic Way'

Those who can freely wield the *tachi* in accordance with the Way are known as strategists. On the path of martial arts, those who can shoot well with the bow are called archers; those who use a gun are called musketeers; spear users are called lancers, and *naginata* wielders are called pike wielders. However, we do not call *tachi* or *wakizashi* users *tachi*-ists or *wakazashi*-ists. Bows, guns, spears and *naginatas* are all weapons used by warriors, and each has a place in the Strategic Way.

Even so, there is a reason we refer to the *tachi* alone in the Strategic Way. By virtue of the *tachi*, warriors govern the world and themselves, and so the *tachi* is what strategy

originates from. Someone who comprehends the virtues of the *tachi* can beat one, or even ten opponents. And if one person can beat ten, then one hundred people can beat one thousand, and one thousand people can certainly beat ten thousand. In my *Niten-Ichiryū*, defeating one person is the same as defeating ten thousand; which is to say that the strategy of *Niten-Ichiryū* encompasses all parts of the Warrior's Way.

With further consideration of the concept of the Way, those who follow the Ways of Confucianism, Buddhism, tea ceremony, tutelage of etiquette, or traditional *Nōh* dance are on different paths to that of the warrior. But even those on different paths will discover commonality with each other if they have a broad knowledge of their Way. It is important for every single person to achieve excellence in their Way.

7. Understanding the Advantages of Different Weapons in the Strategic Way

If you can fully grasp the benefits of different weapons used in varying combat situations then each one will become a useful tool to you, usable at the right moment.

The *wakizashi* is incredibly useful in a tight spot, or when you want to close in on an enemy. The *tachi* is generally useful in any situation.

The *naginata* falls short of the spear on a battlefield. The spear is offensive, whereas the *naginata* is more defensive. In the hands of two equally skilled warriors, the spear is a little stronger. However, depending on the circumstances,

A fortuneteller (left) examines the physiognomy of famous swordsman Miyamoto Musashi (right).

A mounted samurai fully equipped with bow, arrows and katana *sword, as well* kabuto *(helmet),* sode *(shoulder armour),* haidate *(thigh armour) and* suneate *(shin guards).*

both the spear and the *naginata* have their limits in a confined space. They are both similarly limited against opponents entrenched in a defensible position. They are solely weapons for the field, where they are terribly effective in battle. But if you focus on using the weapons solely indoors and get swept up in their intricacies, you risk forgetting their true Way and rendering them useless to you on a real battlefield.

The bow can be used in tactical manoeuvres against enemies, alongside a company of spear or other weapon users. Arrows can be fired with great speed so bows are of particularly good use on an open and relatively flat battlefield. However, it is an insufficient weapon to use for a castle siege or if the enemy is more than 36 metres (118 feet) away. These days, it goes without saying that archery and indeed many of the military arts are for nothing but show, and their actual use is being neglected. 'Arts' like this are not at all useful when it comes down to using them.

Inside a castle is where the gun comes into its own. It also has many advantages out in the field before a battle begins, but once the fighting has commenced it is not as effective. A comparative advantage of the bow is that the arrows fired can be seen by the human eye, so the kills can be counted. Bullets, on the other hand, have the disadvantage of not being traceable with the naked eye. Think hard about this.

It is important for horses to be responsive to reins and be well disciplined. In general, any tool of war should be

up to the task. Your horse's walk, the cutting edge of your *tachi* or *wakizashi*, the stabbing point of your spear or *naginata*, your bow and gun should all be strong and unbreakable.

It is also unacceptable to have a particular preference for weapons. Too much of something is as bad as too little. Instead of copying another's weapon choice, choose one that suits your mindset and feels right in your hand. It is wrong for generals and soldiers to harbour preferences for particular things. Working on this is important.

8. The Rhythm of the Way

All things have their own rhythm, and the Way of the Warrior is no different. The rhythm will not become clear without finely tuned training. Each rhythm of the world is perfectly in time with its art. *Nōh* artists, musicians, instrumentalists, etc., all have their own, fitting rhythm. There is a tempo to everything in the Way of martial arts, whether it's firing an arrow, shooting a gun or riding a horse.

In all the various performing and craft arts, one must keep to the rhythm. There are even formless rhythms; rhythms to the things we feel rather than see. Over the course of a warrior's life, there is rhythm in serving one's lord well, and rhythm in the disgrace of exile. There is rhythm in a harmonious life and rhythm in a discordant one. Even in the path of the merchant, there is rhythm in riches and rhythm in ruin. There is a beat to every path. You must discern for yourself the rhythm of prosperity and the rhythm of decline in all things.

Performed from the tenth century at court functions by nobles and professional musicians, the bugaku *dance was an important part of life at the Kyoto imperial court.*

Kendo training was a key method for samurai to hone their swordsmanship skills.

There is so much that makes up the rhythm of the Strategic Way. First, the rhythms of harmony and discord with your opponent. Then among the dynamics in speed and volume, the rhythms of striking, pausing and countering. Knowing all these rhythms is crucial for the Strategic Way. Without an understanding of the rhythm to counter your opponent, the Strategic Way will not resound within you.

Unravel the rhythms of your opponent and catch them off-beat with one of your own. Use your understanding to find formless rhythms, and your wisdom of the Way to take victory. I write about the importance of these rhythms throughout the scrolls. Consider everything I record and use it in training.

9. *Following the Path*

With consistent training from morning till night, your mind will open itself to my school's Strategic Way of the Warrior. I convey below in writing, in the five scrolls of Earth, Water, Fire, Wind and Expanse, the Way of the Warrior for both the individual, and the masses throughout the world.

For those who intend to follow my Strategic Way, commit yourself to the path by following these rules:
1. Do not entertain corrupt thoughts.
2. Forge yourself by training in the Way.
3. Familiarize yourself in a wide variety of arts.
4. Understand the Ways of all professions.
5. Discern the integrity of all things.
6. Perceive the true value in all things.

A daimyo *(warlord)* watches samurai practising their swordplay in this 19th century illustration.

7. Realize that which the eye cannot see.
8. Be attentive to even the minute details.
9. Make sure that everything you do has purpose.

These are the principles you should keep at the front of your mind as you train in the Way of the Warrior.

Mastery of this particular path is extremely difficult if one cannot see the full extent of what the fundamentals stand for. However, if you can learn these rules then you will never lose to twenty or even thirty opponents, even single-handed. If you set yourself with wholehearted determination upon the Strategic Way and walk the direct path, you will always have the upper hand, even being able to overpower people with just the look in your eye. Your body will learn to move freely as you take it through training, bringing with it a strength to overpower others physically. Finally, with your spirit fully in tune with the path, you can overpower people with your strength of mind. Once you have made it this far, what could your enemy possibly do to faze you?

In the bigger picture of the Strategic Way, for a general commanding soldiers, it is about having the edge with superior troops, having the edge when manoeuvring large numbers of warriors, having the edge in self-discipline, in governing one's province, in providing for the people, and in upholding the laws of the world. No matter what path you take, knowing how to avoid losing to others and making a name for yourself is key. That is the Strategic Way of the Warrior.

By Shinmen Musashi-no-Kami-Genshin, 12 May 1645

To my disciple, Terao Magonojō

水の巻

The Scroll of Water

The essence of *Niten-Ichiryū* roots itself in water, which is why I record the *Ichiryū* way to wield a sword as the Scroll of Water.

It is difficult to explain the intricacies of this path exactly as they are in my head. But even if my words are not enough to explain it, sufficient contemplation should give you an intuitive understanding. Every word, every phrase in this book requires careful reflection and consideration. Inconsistent thought will result in misled notions about the Way of the Warrior.

Although the principles I write about for victory in the Strategic Way are for battles between individuals, it is important to also perceive them on the larger scale and understand their use in battles between thousands. This path is finicky, in that even a slight misjudgement will lead you in the wrong direction and onto a contrary, tangential path.

This print by Kunisada Utagawa (1786–1864) shows a daimyo *(warlord) contemplating military strategy.*

You will not be able to understand the Way of the Warrior by simply reading this text. Take all of the concepts recorded in this work as though they are written explicitly for you. Do not think to just glance over them, do not think to simply learn, nor copy them. Rather, believe that they are concepts flowing from your own heart that you are striving to embody with complete dedication.

1. *The Mindset of the Strategic Way*

Those upon the path of the Strategic Way must maintain a stable mindset; one that remains unchanged from day to day. In both everyday life and in strategic training, your mindset must remain steady. Keep an open and directed mind, free of tension, yet not at all lax. Keep your mind from wandering, keep it centred, but keep it calm and flexible. This is something you must contemplate carefully so that your mind can go with the flow, rather than be swept away by it.

The mind should not idle in times of tranquillity, nor should it rush, no matter how fast things move. The mind should not be subservient to the body and the body should not be subservient to the mind. Give thought to the movement of the mind, but do not obsess over the movement of the body. Your mind should not be insufficient, but equally should not be a drop in excess. Even if your surface thoughts are not behind iron defences, make sure your innermost thoughts are protected from people who might try to see through you.

A smaller person must keep in mind that which is larger than themselves, and a larger person that which

is smaller, and it is important for both large and small people to remain impassive, making sure they are not tricked into thinking with preconceived notions of the implications of size.

Keep your mind clear and open, and you will have plenty of space for all your knowledge. Polishing both mind and wisdom is of utmost importance. Enriching your wisdom will enable you to discern between the logical and illogical in this world and know what is right and wrong. Follow the paths of every art far and wide, and eventually you will become undeceivable by anyone that walks the earth. Then, and only then, will the first real understanding of the Way of the Warrior bloom in you.

Wisdom in the Way of the Warrior is distinctive. On the battlefield, when you are engaged in a thousand things at once, remember to concentrate on the principles of the Strategic Way and find the immovable mindset within you.

2. *One's Bearing in the Strategic Way*

Your posture is important. Face straight ahead, without tilting or leaning your face to the side. There should be composure in your eyes, but do not wrinkle your forehead. Furrow your brow slightly and keep your eyes steady, unblinking and slightly narrowed, with a wide view of your surroundings. Think of holding your head high with your chin sticking out slightly. Keep your neck straight and strong, shoulders down and back. Imagine your body as one line from the shoulders down, keeping the back straight and the pelvis tucked in. Stand strong

'A samurai and the Conquered' by Utagawa Kuniyoshi (1798–1861).

Armed with two swords, the actor Ichikawa Danjuro V poses as a samurai ready to fight (circa 1771).

through the knees and feet, keeping your stomach tensed to stop your lower back from arching. The final piece that brings it together is the *wakizashi*. Sheath the sword and, keeping it pressed against your stomach, make sure that your belt is not slack. This is called 'tying the linchpin'.

As a general rule, for good posture according to the Way of the Warrior, it is important to make your everyday bearing your battle posture, and your battle posture should become your everyday bearing. Keep this in mind.

3. One's Gaze in the Strategic Way

Your gaze should reach far and wide. 'Looking' and 'seeing' are two different things. The gaze that 'sees' is intense, whereas the gaze that 'looks' is soft. Being able to see distant things as if they were close, and close things as though they were far away, is of the essence for the Strategic Way. What I mean by this is that you should know exactly where the enemy's *tachi* is, but not spare even a glance to look at it. This is essential in combat and you must strive for this ability.

Whether the battle is one on one, or a full-out pitched battle of many people, your gaze should be the same. Watching what is in your peripheral vision without moving your eyes to look to either side is critical. This is not something that can be done suddenly in the heat of battle. This is something you must work on constantly to make sure that you always remember what is written in this scroll and are able to maintain the same gaze at all times, no matter what happens.

観と見との二つの眼

4. *The Tachi Grip*

To grip the *tachi*, hold it very lightly with your thumb and forefinger, and with your middle finger neither tight nor loose. Finally, your ring and little fingers should close tightly upon the hilt. A slack hand is unacceptable. The *tachi* is for cutting down your opponent and you must wield the sword accordingly.

As you cut your opponent, your grip should not change on the hilt, and your hand should not tense up. The only thing that should change slightly as you slap, parry or press the enemy's sword is your thumb and forefinger. Regardless, you should still wield the sword with the intent to cut. This does not change whatever the reason you hold the sword; be it a test swing of a new blade, a strategic cut or a killing blow.

A rigid hand or a rigid sword is inadmissible. A rigid hand is a dead hand. A fluid hand is alive with motion. Take this to heart.

5. *Footwork*

When moving your feet, stay light on your toes and press strongly through your heels. Depending on the circumstances you might use big or small, fast or slow steps, but you should always walk as normal. The three movements known as jumping feet, floating feet and heavy feet are terrible for footwork. Instead, *Yin and Yang* footwork is important upon this path. By this I mean never moving just one foot at a time. When you're cutting, pulling back or parrying, move your feet as *Yin and Yang*; right and left, right and left. There should

Rikaku II, a popular Osaka actor, is shown gripping the hilt of his sword at a climactic moment in the vendetta play, Revenge in Front of the Palace.

never be a time when you move just one foot. This is something you must deliberate.

6. The Five Stances

The five ways to stand are: high stance, middle stance, low stance, left-side stance and right-side stance. Even though the stances are classified in five ways, they are all for the purpose of cutting an opponent. There are only these five stances, but every single one should be in your mind for cutting, not for posing.

Assuming a broad or narrow stance is dependent on the circumstances and whether it will be to your advantage. The high, middle and low stances are the basic positions, while the left- and right-side stances are for more applied situations, such as when the ceiling overhead or your flank are obstructed. Use of the left-side stance and the right-side stance should be judged according to the battlefield.

You must understand that on this path, the middle stance is the most important; it is the cornerstone, the foundation of all stances. If you observe the Strategic Way as a whole, you can see clearly that middle stance is the army general, and the four remaining stances follow the general's lead. Always keep this in mind.

7. The Sword's True Path

Truly understanding the Way of the *tachi* means knowing the path that your constant companion takes through the air, and knowing it well enough that you can swing it freely, even if you hold it in just two fingers. Trying to swing the *tachi* too fast will make it harder to wield properly along the

'The First Nakamura Nakazo as a Samurai' (1770), by artist Katsukawa Shunshō.

This 18th century print by Toyokuni Utagawa shows a samurai weilding a flaming torch in his left hand while his right hand rests on the hilt of his sword.

path it should take. Instead, swing the *tachi* with composure so that it is effortless to wield. You cannot swing the sword along the correct path if you brandish it frantically, as you might a folding fan or a dagger. Doing so is known as stabbing like a 'sewing machine' and will not work for actually cutting someone with a *tachi*.

After you swing a downward blow, raise the sword back up along the same path. When swinging sideways, return the sword along the same horizontal. In essence, use the full length of your arm for a heavy swing. That is the Way of the *tachi*.

If you master the five extrinsic forms of my *Ichiryū* School, the swing of your sword will effortlessly follow the path it is designed to take. Train with persistence.

8. The Five Extrinsic Sword Forms: Number One

Middle stance is the foremost stance. From this stance, meet your opponent with your sword tips already in their face, ready to deflect the oncoming attack to the right, with your sword riding on top of theirs. When your opponent comes in for a second attack, retaliate with a downward blow of your blade to knock their sword down and keep it there. If they attack a third time, strike at their arms from below. This is the first form.

It is difficult to comprehend the Five Forms just from what I write here. These Five Extrinsic Sword Forms are recorded so that you might take up your *tachi* and train to swing it along its path. With these five forms you will

A samurai poses in a high stance with raised sword, circa 1860.

come to understand your sword's trajectory and be able to predict any and all incoming attacks from your opponent. *Niten-Ichiryū* has no other forms than these five, and you must build experience in each of them.

9. The Five Extrinsic Sword Forms: Number Two

High stance is the second of the stances. Just as your opponent comes in for the attack, strike a single blow from high stance. Keep your sword on top of the enemy's deflected blade unless they attempt another attack, in which case you can sweep your blade upwards. It is the same no matter how many times they attempt their attack.

The five forms each have their own mentalities and rhythms. If you build up experience in *Niten-Ichiryū* with the Five Sword Forms you will understand the path of your sword so intimately that you will make no blunders and you will come out victorious no matter what situation you find yourself in. Building on your experience is imperative.

10. The Five Extrinsic Sword Forms: Number Three

The third of the stances assumes the low stance, with your swords pointing down and your spirit grounded. When the enemy comes in to attack, swipe upwards towards their hands. If the enemy attempts to strike again after your upwards swipe, keep the same rhythm by striking at them once more, cutting across their upper arms with a horizontal swing. However, from low stance you should aim to stop the enemy's blade with one strike.

Wielding your *tachi* along its rightful path in low stance allows the user to adapt to the opponent's movements, whether they are fast or slow. Take your sword in hand and experience this for yourself.

11. The Five Extrinsic Sword Forms: Number Four

In the fourth stance, begin with the swords at your left side. Strike at your opponent's hands from below as they reach forward to attack you. If the enemy makes an attempt to knock your upswinging blades down, concentrate on cutting their hands and continue your swing along its path, diagonally upwards to above your own shoulders. This is the path of the *tachi*. If the enemy comes in for another attack, use the correct trajectory of the sword to be in the winning position. This warrants deep consideration.

12. The Five Extrinsic Sword Forms: Number Five

In the fifth stance, begin with the swords held horizontally at your right side. Parry the enemy's incoming attack, and bring your *tachi* diagonally up to transition to high stance from low stance. Follow through with a straight cut downwards. This is fundamental for understanding the path of the *tachi*.

By becoming familiar with swinging your sword in the fundamental stances you will be able to wield the heavy *tachi* with incredible freedom. I won't go into great detail about the Five Extrinsic Sword Forms in writing. However, you should know that through an all-consuming dedication to practice of the Five Sword Forms, you will gain an

A samurai poses in a low stance, katana *at the ready.*

understanding of swordwork in *Niten-Ichiryū*, and the ability to embody the ensemble of rhythms and read the movements of your opponent's sword.

Whatever you might come up against, you are guaranteed victory in battle by honing your skills with an infinite amount of sword training, pre-empting anything your opponent might try, and keeping tempo with various rhythms. Research and reflect upon this.

13. The 'Stanceless Stance' Teaching

What I mean by this is that these positions for wielding your sword should not be defined, lest we fixate on them. But nevertheless, those are the five ways to hold the sword that I define as stances.

The way you wield your sword depends on the opponent, the place, the circumstances, and even the direction your enemy chooses to face, but you always wield it so that it is easy to cut down the enemy. Sometimes in high stance you'll have the mind to come down into middle stance, and when the moment strikes it might be beneficial to shift back up into high stance. From low stance you might have the chance to come up to middle stance. Even in the side stances, the situation might call for you to bring your swords to the centre, bringing you back to middle or low stance. Because of this fluidity, my principle is that there are stances, but no defined stance.

If you have taken up your sword, whatever comes next, you are ready to cut your opponents. You must be absolutely ready to fell every enemy, whether you parry, slap, hit, press or just lightly tap the sword being brandished in your face. If you concentrate on merely parrying your opponent's sword, or just slapping it, hitting, pressing or tapping it, you have forgone the ability to cut your enemy. No matter what movement you perform, it is essential that you regard everything you do as solely for the purpose of cutting your opponent. Think on this carefully.

The placement of troops in battle is equivalent to a stance, and all stances are a formation by which to win a battle. It is not good to get set in your ways, fixating on any one formation you know well. Take this into consideration.

14. Striking Your Opponent in a Single Beat

To strike your opponent in a 'single beat', you must move into striking distance before your opponent even registers your imminent attack. The single-beat strike is so fast and true that you reveal no movement and your mind works on instinct.

Striking before your enemy has the chance to decide whether they will draw their swords, dodge or attack is the essence of the single-beat strike. Once you have built the muscle memory for this, you can work on training yourself to take advantage of the off-beats and strike at speed in the pauses between rhythms.

15. The Double-Beat Rhythm

The double beat is to be used to outmanoeuvre your opponent's hasty parry of, or retreat from, your attack.

A ronin, *or masterless samurai, lunges forwards in this 19th century illustration by Yoshitoshi Taiso.*

Start with a feinted strike, and once your opponent feels secure in their parry or dodge of your feint, catch them off-guard with another strike. This constitutes the rhythm of the double beat.

It will be almost impossible to pull this off after merely reading the theory, but if you take your sword in hand and receive proper training you will swiftly appreciate the idea.

16. Being 'Unconscious, Unpredictable'

In moments when you strike at the same time as your opponent, your unconscious body and mind take over, and all they know is that they have to hit the enemy. Your hands will strike of their own accord with a power and rapidity that can only come from this ability called the 'unconscious, unpredictable' strike, and it is incredibly important.

This strike can be used against your opponent many times over. Build this into your muscle memory, embody it, and continue to build upon it.

17. The Flowing Water Strike

Use this strike when you are face to face with the opponent you are fighting and they attempt to quickly retreat, dodge or disengage from your sword. Allow your body and your spirit to swell, moving as slowly and steadily as stagnating water. Your sword should follow your body in a large and powerful strike. With sufficient training, this is simple yet effective. It requires thorough scrutiny of the enemy's posture.

無念無相

18. Opportunity Strikes

When you unleash an attack and the enemy attempts to block or deflect it, use the opportunity to cut at their head, hands and legs in one strike. Striking at everything along a single *tachi* pathway is what I call 'opportunity strikes'.

It is best to practise this thoroughly, as it has countless practical applications against your enemy. You should learn this through innumerable repetitions in combat.

19. Striking Flint and Steel

When your sword is locked with your enemy's blade, strike as hard and fast as you would a flint and steel without raising your *tachi* even the slightest amount. Engage your legs, torso and arms in unison to strike in a flash.

This move is difficult to pull off without repeated practice. Keep working on it to increase the speed of the strike.

20. Autumn Leaves Strike

This move is designed to disarm your opponent with your strike, their sword falling to the ground.

As your opponent assumes a stance to strike, slap or parry you, counter with a powerful 'unconscious, unpredictable' strike, or 'flint and steel' strike to your opponent's *tachi*. Force the point of your blade down and keep pressing the opponent's sword with all your mental and physical might and they will eventually drop their weapon.

Two samurais fighting with swords on the roof of a tower.

This 18th century illustration shows two actors involved in combat, the standing samurai about to strike the sitting figure with his sword.

With training, striking down the enemy's sword and making it fall from their grasp is easily done. Make sure to train well.

21. Body Over Sword

This is also known as 'sword over body'. Usually when taking position to strike your opponent you do not move your sword and body at the same time. Depending on your opponent's intent, you can take up a striking position with your body first, and the sword will follow to land a blow. Occasionally there might be times where you strike with the sword first without facing your enemy, but as a rule the body moves to strike first and the sword follows through after. Think about this carefully as you learn to strike.

22. The 'Strike' and the 'Hit'

To Strike and to Hit are two different things. Having the mind to strike means to follow through completely with the blow, no matter the form. A hit, on the other hand, is similar to having a mind to test the waters. No matter how hard you hit, or even if the opponent dies, it is still just a hit. A strike is a blow that only comes from the mindset to follow through. Explore the differences.

To hit the arms or legs of the enemy is just the first step. It should be followed by a critical strike. To hit is equivalent to merely touching. With training, you will come to see that striking is fundamentally different. Think this over carefully.

This colourful artwork captures the confusion and rapid movement of close combat as two samurai fight.

23. Embodiment of the Autumn Monkey

To embody the autumn monkey means to adopt a form whereby you keep your arms close to your body. Think of nothing other than keeping your arms in when you advance into the enemy's space. Before your enemy attacks, get into their space as swiftly as you can and stay there. If you extend your arms, your body will be lagging behind them, so instead you want to be in the mindset of moving your entire body as fast as you can. Getting within arm's length makes it much easier to swoop in on the enemy. Consider this carefully.

24. Embodiment of Lacquer and Glue

The nature of the lacquer and glue idea is to stick so closely to the enemy that they can't shake you. Stay so close to them that you cover them head to toe and are practically inseparable. People tend to move first with their head or legs, leaving their body to lag behind. Make sure to attach yourself to your enemy so closely there isn't any space between you. This bears thinking about closely.

25. A Measure of Height

Whatever the situation, when you are within arm's length of the enemy, you must take care to make your stature greater than theirs. Lengthen through the legs, elongate your spine, crane your neck, and stand strong, face to face with your opponent. It is imperative that you stand tall, stretching so much that you tower over the enemy. Think about this well.

26. Adhering Blades

When your opponent strikes and you strike at the same time, if your opponent parries, encroach on their space with a mind to bind your *tachi* together. Push into their space with just enough calm determination to keep the swords together firmly, but not with too much power. Once you have bound your sword to the opponent's, you can enter their space without the need to rush.

There's 'adhering' and there's 'entangling'. Adhering is a strong move, but getting entangled is weak. You should know the difference.

27. Body Slamming

When you have got into the enemy's space, use the body slam to smash into them with your whole weight. Turn your head to the side slightly and slam your left shoulder into the enemy's chest. Use the full strength of your body to smash into them, moving to the rhythm as if exchanging blows, ready to rebound off the impact. If you practise getting into your opponent's space, you can knock them flying ten or twenty feet (three or six metres), possibly enough to kill them. Build your strength in this and train hard.

28. The Three Parries

There are three kinds of parry.

The first is to parry the enemy's *tachi* as it comes towards you when you advance. Deflect the enemy's *tachi* over your right shoulder and stab your other sword towards their eye.

This print shows the cut and thrust of combat, as fighting samurai seek to strike the fatal blow.

This action-packed 14th-century fragment is from a handscroll illustrating the epic narrative The Tale of the Heiji Rebellion, *which describes the confrontation in late 1159 between two military clans: the Minamoto and the Taira. This detail from a battle scene depicts a frantic melée, with two warriors of the Taira clan closing in on a Minamoto soldier.*

Second, while stabbing in the direction of the enemy's right eye, repel their sword by following with yours towards their neck.

Third, advance upon the enemy's space with your short sword leading, all but ignoring the long *tachi* that you parry. Think of the follow-through with the sword in your left hand as a punch to the face.

These are the three parries. Think of making a fist and hitting their face with a left-handed punch. Drill this continually for experience.

29. Stabbing the Face

An important thing to always have in mind is that when engaged in a fight, you should be aiming to stab the opponent's face with the tip of your sword. If you are committed to stabbing them in the face, your enemy will be forced to duck or withdraw. When your enemy is on the retreat, you will see numerous openings for a winning thrust. Think this over.

If you can see an opening as your enemy evades during the fight, you have already won. Never forget to stab at the face. You must build experience in this as you train in the Way of the Warrior.

30. Stabbing the Chest

You can stab the enemy in the chest when the environment of your fight is restricted overhead or the sides are blocked, making it difficult to cut. To deflect your opponent's

sword, face the blunt of your blade to the opponent and, making sure your sword is not slanted, thrust the tip towards the chest. Make use of this technique when you are tired, or when your blade is blunter than it should be. Run this technique through repeatedly.

31. The 'Lash-Slash' Stab and Cut

Use the 'lash-slash' riposte when your opponent attempts a counterattack after you have advanced on them, forcing them into retreat. Raise your sword from below as if to stab them, and then change your thrust to a strike. Keep doing this to the rhythm; a snappy lash, slash. Thrust an upward lash, strike a cutting slash.

This back-and-forth beat can be used in any battle, no matter the circumstances. Find your lash-slash rhythm by setting your mind on stabbing the enemy with the point of your sword, and then striking them using the momentum created by raising the sword. Practise this even as you consider it.

32. The Parrying Sword Smack

This technique is to be used when you find yourself engaged in a stalemate of clashing swords. Interrupt the enemy's strike with a slap of your *tachi*, followed by a strike.

The interrupting slap does not need great force, but nor should it be a simple parry. Smack the enemy's oncoming *tachi* with yours and follow with an instantaneous strike.

Taking the initiative on the parrying smack and keeping it for the strike is crucial. If you can keep to the beat of this technique, your sword will stay steady in your hand as you fend off any strike from your opponent, no matter how powerful, with just a light smack. Consider this and drill it well.

33. *Engaging Multiple Opponents*

When you are faced with multiple enemies at once, draw both your *tachi* and *wakizashi,* and take a broad, side-to-side stance with your swords out as if to fling them left and right. If the enemy comes at you from four directions, push them back towards one. When they come at you, discern the order of their advance and deal with the closest first. Keep a broad view of your surroundings and when you know where your opponents will strike from, swing both swords simultaneously, cutting the enemy in front as you strike out, and felling those at the sides as you return your blades. Do not wait between swings. Immediately resume your broad stance and cut into the advancing sides. Tear through the enemy and keep charging towards their position, cutting them down as you move.

Find a way to force the enemy into single file. As soon as they are lined up like a string of freshly caught fish, slice into them without missing a beat. Recklessly chasing after smaller groups will work against you. Furthermore, knowing where the enemy will come from but waiting for them to come to you will also work against you. Victory will

Morozumi Bungo no kami Masakiyo, *from the series 'Courageous Generals of Kai and Echigo Provinces: The Twenty-four Generals of the Takeda Clan', circa 1849.*

come from moving to the enemy's rhythm and knowing their breaking points.

When you can, train by gathering together multiple opponents and forcing them back. Once you get the hang of it, dispatching one, ten, or even twenty enemies will become easy. Investigate this through reiterative training.

34. The Principles of Engagement

Victory in a sword fight is attained by adhering to the principles of engagement. I cannot record every painstaking detail. It is up to you to learn what it takes to win through discipline and practice. The overall truth of the Way of the Warrior is revealed in sword technique and the details can only be taught in person.

35. A Single Blow

This 'single blow' is the way to undeniable victory. However, it is impossible to grasp the concept without meticulous study of the Strategic Way. But train diligently to strike down your enemy in one blow, and you will find yourself attuned to the Way of the Warrior, freely able to realize the path to victory. You absolutely must hone your skills in this technique.

36. Direct Connection

A direct connection to the deepest level of teaching is what I convey to someone who is fully on the true path of *Niten-Ichiryū*. Temper your body through training to forge true mastery in the *Niten-Ichiryū* Strategic Way of the Warrior. This is important. Other details must be taught to you orally.

What I have recorded above amounts to the main substance of my school of *Ichiryū*.

To learn how to win once you have taken your sword in arms against someone, the first things you must learn are the Five Stances with their Five Extrinsic Forms. Once you also grasp the Sword's True Path, your movement will be unhindered and your dynamics will be second nature. You will learn to feel the rhythm in striking and your competency in sword technique will become ingrained. As you start being able to move your legs and body in exactly the way you intend, you will defeat one person, two people, and eventually come to understand strengths and weaknesses in strategy.

Study the articles of this scroll one by one as you take on opponents and gradually get to grips with the principles for victory. Set your heart on lifelong discipline and endless patience, seeking actual experience in the field that will teach you the virtue of it all. Engage any and all kinds of people and learn their minds. Walk the thousand-mile path, one step at a time.

The Way of the Warrior requires unhurried and composed thinking in order to forge an understanding of the path. Become attached to the Strategic Way as the samurai's calling. Today's victory is being better than the you of yesterday; tomorrow's victory is conquering your weaknesses; and building on your strengths is your victory for the future. Keep in mind everything written in this work, and do not let your mind wander off-course even slightly. Even if you win against the most formidable

The celebrated actor Ichikawa Ōmezō (1781–1833) is shown in a formal pose and garbed in a persimmon brown ceremonial kimono.

enemy, but you do so by defying the teachings, you are not following the true Strategic Way of the Warrior.

The knowledge to defeat tens of men by yourself comes from incorporating the principles for victory in all that you do. The power and wisdom you accrue in your swordwork will make way for mastery of the Strategic Way in all-out battles or one-on-one combat.

Forged over one thousand days of training, sharpened over ten thousand more, and you are close to being called tempered. Scrutinize these words.

By Shinmen Musashi-no-Kami-Genshin, 12 May 1645
To my disciple, Terao Magonojō

Entitled 'Theatre Scene' (1844), this print shows samurai involved in an all-out battle to the death.

千日の稽古を鍛とし、
万日の稽古を練とする

火の巻

The Scroll of Fire

Introduction

Battle is like a raging fire in the *Nitō-Ichiryū* Strategic Way. I will shed light on all things contest in this Scroll.

The people of the world have a narrow view on the principles of victory. Some focus on using nothing but the movements of the wrist and fingers to win, while others think of waving a fan back and forth, focusing on the forearms. Still others place emphasis on just a little extra speed for weapons like the bamboo sword, or work on the agility of the hands and feet.

To keep following the Strategic Way, I have put my life on the line in countless contests. I have gambled with life and death to learn the Way of the sword, I know the strengths and weaknesses of an enemy's *tachi* as they strike, and I know the proper use of the edge and back of my blade. To be tempered as a person ready to slay an enemy, it does

not do to dwell on small, weak matters. Particularly when you are in a full suit of armour, it is no use concentrating on the little things for just a slight advantage in battle. The fully realized meaning of my Strategic Way is going alone up against five, or even ten opponents in a fight for your life, but knowing the path to victory without a shadow of a doubt.

With that as our premise, is there any difference in one person beating ten, and one thousand people beating ten thousand? This is something to consider. Even so, it is impossible to amass a thousand or ten thousand people to train with. This means that as you take up your sword alone, every single time you must weigh up the enemy's resourcefulness, know their tricks, strengths and weaknesses, and employ your knowledge of strategy. This is how you come to a place of mastery where you are able to beat thousands of people.

Tell yourself, 'Who else in this world but me can access the Strategic Way of the Warrior?' and 'Which of us is the best person here?'. Use these thoughts to train assiduously from morning to night. Once you are forged, tempered and polished, you will achieve a freedom like no other. You will discover inexplicable power; power that will be useful to you in any situation. This is the spirit that a warrior needs to employ clever strategy.

1. The Situation at Hand

There is a concept called 'shouldering the sun' for assessing the battlefield. Put simply, it is to stand with

Dating from the 1870s, this photograph shows a samurai dressed in haramaki *style armour, armed with both the long* katana *and short* wakizashi *swords.*

your back to the sun to be harder to see. If the area doesn't allow for keeping the sun behind you then keep it on your right side instead to make your *tachi* difficult to see. It is the same as being inside and keeping the light behind or to your right. You want to make sure to keep your back and right side close to a wall or obstacle, and keep your left side free to maintain an accessible sword. This remains true for nighttime if your enemy is visible; get used to keeping the firelight to your back or some illumination to your right. You should also get used to taking what is called the high ground; an area that is at least a little higher than the enemy's position, from which you can look down upon them. When indoors, think of the tallest seat as your high ground.

For the fight itself, it is important to force the enemy round to your left side, giving them no backward room to manoeuvre and driving them into a tight spot. The mindset is to relentlessly press the enemy so that they don't have the time to turn their heads and see the predicament you have driven them into. Even indoors, keep them unaware of the thresholds, lintels, doors, sliding doors or edges that you chase them into.

Whether it's a tight spot, a pillar or an obstacle, it's all the same. However you choose to rush your enemy, make full use of the lay of the land, whether it's uneven footing or an area with no room to the sides. Pursue the advantage of terrain with tenacity. Explore the possibilities here and train hard.

A samurai warrior takes up a high stance. The body position depends on the disposition of the opponent.

Miyamoto Musashi kills a giant serpent using a naginata.

2. The Three Initiatives

There are three ways to take the initiative in battle. The first of these is to make your move before your enemy does. This is called the 'advance initiative'. The next is to attack the enemy after they make their move towards you and is called 'delayed initiative'. The last is called the 'reciprocal initiative' and involves attacking the enemy at the same time as they attack you. These comprise the three initiatives.

No matter what type of fight it is, once the battle begins, these are the only three initiatives that matter. Taking the initiative is the way to a quick victory, and thus a core attribute of the Strategic Way. There are a multitude of specifics for initiatives; you will triumph by using the right initiative principles for the situation at hand in cooperation with your reading of the enemy's intent and your wisdom of strategy. However, I cannot record every single detail in writing.

NUMBER 1. THE ADVANCE INITIATIVE

When you want to be the first to attack, compose yourself and then seize the initiative in one rapid motion. Show your speed and ferocity on the outside, but remain tranquil within. Another way to achieve this initiative is to strengthen your resolve and move your feet faster than you would normally, to get into the enemy's range for a rapid onslaught. Or unleash your spirit to keep crushing the enemy throughout the fight, keeping the same, strong determination to win that you had from the start, right through the fight to the end. These are what advance initiatives truly look like.

NUMBER 2. THE DELAYED INITIATIVE

When the enemy comes at you, give not the slightest reaction and feign weakness. As soon as they come in close, spring away suddenly, with your readiness to strike apparent in your posture. Watch for the moment the enemy falters in surprise, going in for the kill as they do. This is one way of taking the delayed initiative. Alternatively, as your enemy attacks and you throw out even more strength to overcome them, you can take advantage of the moment between the changing rhythms of attack to secure your victory. This is the essence of the delayed initiative.

NUMBER 3. THE RECIPROCAL INITIATIVE

When the enemy comes in with a hasty attack, remain composed and ready your assault. Once they get close, lash out without warning. Move so suddenly that you'll have won while your enemy is still crawling at a snail's pace. However, if your opponent comes in for a more composed attack, get light on your feet and ready a swift assault. Use the enemy's movement against them if they get in close enough to bash into you and follow the momentum through for a powerful finishing blow. These are the reciprocal initiatives.

Writing about the initiatives in clear detail is not easy, so you should contemplate what I have written to understand the concepts yourself.

You won't always be the one to initiate an attack, but when you do, use the three initiatives in harmony with the

In this dynamic woodblock print, samurai Matsumoto Koshiro leaps through a hole in a wall with his sword in mid-swing above his head.

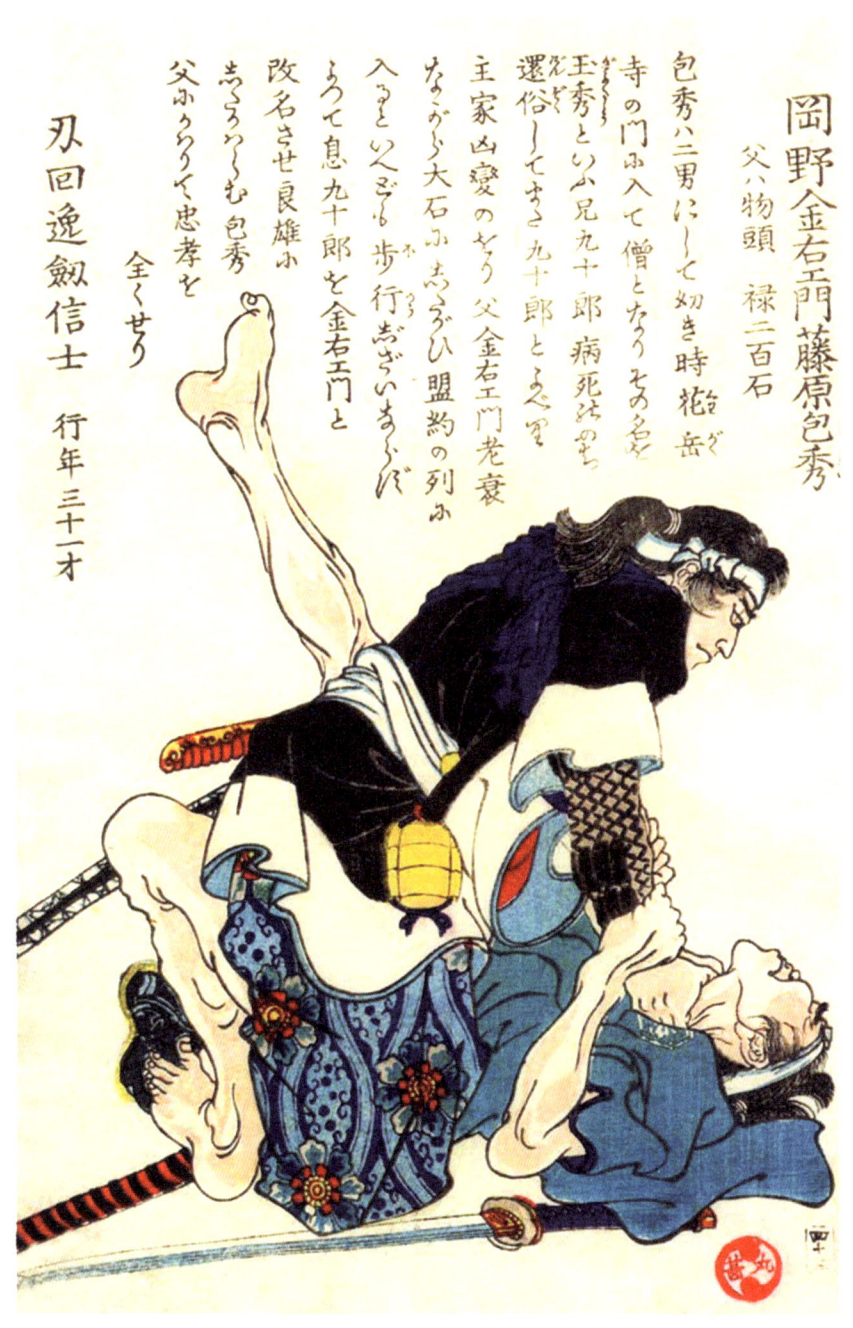

Holding an opponent down, from a series of illustrations depicting 'The revenge of the Forty-seven Ronin', Japan's greatest national legend.

situation and principles at hand, and take that initiative to dominate your opponent's movements. These initiatives require thorough training to perfect the spirit of certain victory that comes with the wisdom of the Strategic Way.

3. Nip it in the Bud

Nipping it in the bud means keeping the enemy's head down, as though you are smothering them with a pillow. The Strategic Way dictates that in contests it is dangerous to be controlled by the enemy, or be a step behind them. Whatever it takes, you want to be the one with total control over your opponent's movements. Of course, your opponent will be thinking the same thing, so breaking their movement patterns won't go as well as you plan unless you can overwhelm them. Blocking as the enemy attacks, stopping thrusts, or twisting free of a hold means you are already behind in terms of the Strategic Way.

Nipping it in the bud means using your experience of the Strategic Way to read the signals your opponent makes in a fight, and stopping them before they even have the chance to make their move. Nip their strike at the letter 's' so they can't continue. That's the mindset for nipping it in the bud. Nip the 'a' off their attack; nip the 'l' off their leap; nip the 'c' off their cut; it's all one and the same.

However, even as your enemy advances, you should be allowing ineffectual techniques to pass, but nipping real ones in the bud to prevent their execution. This is foremost in the Strategic Way. However, be mindful that focusing on suppressing your opponent's attempts over and over will eventually lose you the initiative.

A Japanese wooden ship is rowed out to sea with Mount Fuji prominent in the background.

First and foremost is that any of your techniques, regardless of the context, should be justifiable. If your technique has integrity you can nip any tricks that your opponent might attempt in the bud. One who can keep the enemy in the palm of their hand without allowing them a single useful move is a master of the Strategic Way, and a product of hard training. Investigate the concept of nipping it in the bud as thoroughly as possible.

4. Traversing Pivotal Points

To take the ocean as an example for this, traversing pivotal points means navigating narrow channels, or negotiating hundred-mile-long (160-kilometre) straits. Overcoming such perilous passes is what I mean by 'traversing pivotal points'.

Throughout life there will be many difficult junctions to cross. To sail, you need to know the perilous places that need navigating, the capability of your vessel to handle rough waters, and even how auspicious the day will be for you. If you sail without an accompanying vessel then you have to work with the conditions and tack into the wind or ride the tailwinds. If the wind changes, you need to consider rowing inland 10 miles (16 kilometres) or so to land in port. To get through life, one needs the same sort of mindset as sailing a vessel over uncertain waters. Life requires you to carefully consider how to traverse pivotal points with the right priorities.

Knowing how to cross these crucial points in battle is essential to the Strategic Way. Assessing the enemy's position and being aware of your own competence, and

A battle scene from the series 'The Forty-seven Faithful Samurai', 1840s coloured woodblock print by artist Utagawa Yoshitora.

then using these principles to traverse points in battle is akin to a skilled captain sailing a boat over ocean routes. The calm of the mind comes after the storm. Successfully traversing the critical point in battle results in a weak opponent, gives you the initiative, and brings you to a speedy victory. Being able to navigate crucial points is as important in large-scale battle as it is in single combat. Tread carefully.

5. Knowing the Conditions

In terms of large-scale battle, knowing the conditions refers to the state of the enemy's vigour and morale, knowing their intent, taking the location into account, and discerning their mental state. Fight your battles with this knowledge in mind, using it as a principle to decide how to deploy your own troops and seize certain victory, knowing you now have the initiative.

In single combat, it is important to have an understanding of the opponent's school, character, and strengths and weaknesses. Throw the enemy off guard by catching the variations in their tune and the intervals in their rhythm to decide your offence. You will be able to see the conditions of many different things if you foster a superior wisdom. When you can freely use strategy in combat you will be able to read the enemy's mind and unveil a multitude of paths to victory. Be astute in this.

6. Jumping the Gun

This technique is used exclusively in, and is incredibly important to, battle.

限らない

Firstly, in large-scale battles where the enemy uses bows and *tanegashima* musket guns, it will be difficult to penetrate the enemy lines if you attack under the rain of arrows and bullets they have already fired, as they will be re-nocking arrows and loading powder by the time you are close. To jump the gun, rush in for a rapid attack just as they fire their bows and guns. By storming them so quickly, it will make it difficult for them to nock another arrow, and hard to load more gunpowder. For you to jump the gun, you must be ready for whatever the enemy throws your way by catching on to the most reasonable way to win in the situation and getting the jump on the enemy's offence to seize victory.

In one-on-one combat, striking repeatedly after each of the opponent's blows makes for a continual clanging cadence but does not get you anywhere. You can jump the gun here by stamping on the enemy's *tachi* instead. By trampling their blade you will have won at the outset, as long as you make sure they cannot raise their sword for a second attack. This trampling isn't limited to the feet. You should take the opportunity to jump the gun with your whole body, trample them with your spirit, and, of course, hit them with your *tachi* so fast that they don't have the chance to attack a second time.

This is precisely what it means to take the attack initiative in any situation. Although you should be in sync with the enemy's movements, it is not so that you can smash into each other. It is so that you are already prepared for the aftermath. Think on this well.

Samurai warriors do battle in a scene from 'The Forty-seven Faithful Samurai' by artist Utagawa Yoshitora.

A samurai defeats his opponent in this scene from an underwater battle, taken from Taiheiki Eiyu Den *('Heroic Biographies'), dated 1867.*

7. Detecting the Collapse

Nothing escapes deterioration and eventual collapse. Houses fall, the human body deteriorates, the enemy breaks down. It is at the point when things are no longer in tune that destruction befalls them.

In a large-scale battle it is important to catch the faint notes of deterioration in the rhythms of your enemy and drive forward to make sure you don't let the moment get away. If you miss the signs of deterioration then your enemy will have a chance to recover.

You will see the signs of collapse appearing in your enemy as their rhythms start unravelling during individual combat as well. If you are careless and overlook the cracks, your enemy will get back up for a renewed attack and you will have got nowhere. It is important to press your attack firmly to widen the cracks and make sure your enemy cannot come back from the brink of collapse. Be straight and relentless in your offence; batter them so thoroughly that they can't even stand up. Consider the impact of smashing your opponent to pieces. If they are not utterly destroyed, they remain a threat. Be mindful of this.

8. Put Yourself in the Enemy's Shoes

To think from the enemy's perspective we should put ourselves in their position. The world over, we have internalized that a burglar hiding out in the target house is in a strong position. However, if you put yourself in

the burglar's position you'll see that they have actually just burrowed themselves away from a world full of enemies, with no alternative route out. They are like a caged pheasant with a falcon looming over them ready to strike. Work on your understanding of this.

Even in large-scale battles, we have the tendency to think of the enemy as strong and become overly cautious. But if your troops are capable and you understand the principles that will lead to victory over your opponent then you have nothing to worry about.

The same goes for one-on-one combat; put yourself in their shoes. They see before them an embodiment of strategy who is enlightened in principles of victory; a master of the Strategic Way of the Warrior. They will already be anticipating their own defeat. Contemplate what this means for you.

9. Releasing the Four-Handed Grip

When you and your opponent are of the same mind and at somewhat of a stalemate, think of releasing the hold you both have on one another; a four-handed grip. If you realize you have come to this impasse, it is best to immediately throw out the tactics you were using and adopt new strategy to achieve victory.

If you cannot make yourself let go of your four-handed hold on your current strategy, even in large-scale battles you will make little progress and your soldiers will suffer losses. You absolutely must discard your current struggle

Two swordsmen are locked in combat, and a third has taken a tumble, under the watchful eyes of Karaki Masaemon and Honda Naiki. This episode is taken from a popular revenge story – how the son of a murdered samurai tracked the killer throughout Japan and finally confronted him at Iga Pass. The drama, first performed in 1777, was based on a historical incident from the 1630s.

The battle of Nagashino took place in 1575 near Nagashino Castle on the plain of Shitarabara in the Mikawa Province of Japan.

with haste and take up a new tactic that will outwit your enemy.

 This is just as important in one-on-one combat. If you think you are stuck in the four-handed hold, you should immediately change tack in quick response to the enemy's stance and use a new ploy to ensure you win. Think resourcefully.

10. Lifting the Shadow

This technique is how one overcomes the times when reading the enemy's intent is impossible.

 When you cannot fathom the state of the enemy's situation in a large-scale battle, you can feign an all-out offence to tip their hand into showing their ploy. Once their strategy has become clear it is simple enough to apply the right tactics to secure victory.

 Similarly, in one-to-one fighting when the enemy has their *tachi* hidden to the side or back by their current stance, if you feint unexpectedly then they will reveal their intent in the movement of their sword. Once the enemy's tactics are out in the open for all to see, you can embrace the strategy you need, and immediately go for the win. If you are too careless you will lose the rhythm, so study this carefully.

11. Stifling the Shadow

Stifling the shadow of your enemy is the technique to use when you see that their intent is to attack. In large-scale battle, this means pushing back against the enemy as they

In this colourful illustration a samurai vanquishes his adversary.

are about to go on the offence. If you can show the enemy your ability to completely suppress their strategy, they will falter under the power of your strength and change tactics. At that moment you can also change your own side's approach and seize the initiative to win with an open and focused mind.

In one-on-one combat, use your own rhythm to stifle the enemy's techniques at the first sign you see. Ride the enemy's faltering rhythm and turn it into your own victory, using the initiative for your offensive. Think about this carefully.

12. Acting in Infectious Ways

Almost everything is infectious: yawning, tiredness, and even the time of day affects others. When you see that the enemy troops are getting hasty and losing their composure, if you take your time and act as though you haven't noticed, your enemy will follow that lead and let their guard down. As soon as you think your disinterest has infected the enemy, go for a single-minded, lightning-speed attack, and secure unyielding victory.

This is just as important in single combat; take your time and seize the moment when the enemy relaxes for a relentless attack.

There is a similar technique to this called 'intoxication'. Intoxicate their minds with your behaviour and lull them into listlessness, overzealousness or weakness. Practise this infectious behaviour thoroughly.

13. Antagonizing

There are a multitude of things that can upset us. One of them is being in mortal danger. Another is trying to achieve the impossible, while still another is being taken by surprise. This warrants thorough consideration.

It is important to know how to provoke your enemy in large battles. Attack with such ferocity that your enemy chokes on their surprise. You can then organize your troops to move, taking the initiative and victory before the enemy has a chance to recover.

In a one-on-one battle, enrage the enemy by putting on a show of nonchalance, followed rapidly by a forceful attack. It is crucial that you keep the pressure on, pushing the buttons you can see will provoke them. This will lead to your victory. Consider what a distressed enemy means for you.

14. Invoking Fear

Many things invoke fear in us. We become scared when something takes us by surprise. In large-scale battle, there are many means by which to scare the enemy. Incredible noise can make the enemy anxious. Using the illusion of size to deceive can intimidate. And an unexpected attack from the side can incite fear. Latch on to the disrupted rhythms of the enemy's fear and use that to gain the advantage you need to win.

In single combat, make use of your body to incite fear, terrorize them with your *tachi*, and make them cower at the sound of your voice. Make the sudden and

*At the Battle of Awazu in 1184, female samurai (*onna-bugeisha*) Tomoe Gozen (1157–1247) killed Uchida Ieyoshi, earning enduring fame.*

unexpected move, and while your enemy is recoiling, snatch victory as suddenly as you attacked. This is important; study it well.

15. Fusing with the Enemy

Use this technique when you see that you are in close quarters with the enemy, yet also at a complete stalemate. Keep covering them so closely you essentially become one with them. Find your window of opportunity from within the mix.

Fuse with the enemy in both individual combat and on the larger scale when you find that you can't see your victory as separate combatants. Blend with the enemy so well that there is no distinction between you, and use the fray to find your opening, seize your strategy and take your decisive victory. This is an important concept, so take the time to think it over.

16. Chipping Away at the Corners

Going up against an unyielding force head-on will get you nowhere, no matter how hard you push. Instead, hit the points that stick out. In large-scale combat, look for the placement of the enemy troops and launch your assault on the protruding corners that have little protection. If you can weaken a corner of their forces it will have a debilitating effect on the whole unit. Even as that part is weakened, it is important to pay attention to the other points of protrusion to ensure your victory.

In single combat if you can land blows on the protruding areas of the enemy's body and weaken them even a little,

This early 19th century print shows the powerful female samurai Tomoe Gozen on horseback disarming another warrior.

The Battle of Yashima in 1185 resulted in a defeat for the Minamoto clan over their rivals the Taira clan.

their stance will falter and you will win by a long shot. Explore this notion thoroughly and put together the pieces that lead to victory.

17. Flustering the Enemy

To fluster the enemy is to dishearten and confuse them. On the large scale this involves calculating the enemy's next move and using your prowess in the Strategic Way to bewilder them. Make them question 'here or there?', 'this or that?', 'fast or slow?'. Latch on to their chaotic and wandering rhythms for certain victory.

In a one-on-one battle, get a good handle on the circumstances and employ various techniques to confound your opponent. Feint a strike or thrust, or make them think you're about to come in close. Identifying vulnerability as their rhythm descends into disarray is vital for seizing the opportunity for an easy victory. Work this over in your mind thoroughly.

18. The Three Cries

There are three distinct cries, used before, during and after an encounter. Vocalizing at the right point is paramount. Our voice is a vocalization of our energy. We use it to roar at fires, or at the wind and the waves, for it shows the fierce power within us.

The roar used at the beginning of an all-out battle should be as magnificently overwhelming as possible. Then, in the midst of the fight, a low, rumbling growl should come from the pit of your stomach. The yell for victory is mighty and triumphant. These are the three cries.

This 19th century woodcut depicts a samurai army on campaign during the winter months.

In individual combat, pretend to strike at them while yelling an energetic 'Ay!' to shake your enemy into movement, and then follow the cry with a blow of your sword. Then you can announce your victory with a thunderous cry after striking your enemy down. These are the before and after cries.

However, do not vocalize loudly when striking with your sword just for the sake of it. If you yell during the battle, make sure it is a low roar, in time with the rhythm of the fight. Run this through your mind as you study.

19. Interweaving

When the opposing sides clash in the context of a large-scale battle, weave your way through the enemy by finding and assailing a strong point. As soon as it crumbles, move on to the next and the next. Essentially the idea is to shift the target of your attack as you zigzag back and forth.

This is also crucial when fighting as one against many. You are not dividing your attention here and there, but rather fighting one strong point as the other is pushed back. Find a harmonious rhythm to that of your opponents' to sweep back and forth among them, attacking in reaction to their movements. Keep an eye on the state of the enemy and don't hold anything back as you continue your relentless assault to break through their ranks. A clear victory will follow. You can also use interweaving if you want to get in close to a strong enemy in a one-on-one fight.

Interweaving requires a total commitment to moving in among the enemy, without thought to taking a single step in retreat. Consider what this mindset really means.

20. Crushing the Enemy

To crush your enemy, it is critical to see the enemy as weak and yourself as mighty, pulverizing them in your mind. In a large-scale battle, this means looking down upon a small force, or catching the signs of weakness and distraction in a large force and coming down on them hard, crushing them into the very ground. If you are too soft on them, the enemy might recover. Crush them like a bug in the palm of your hand. Study this way of thinking well.

In times of single combat, if your enemy is not up to your skill level, or if they lose their rhythm and start to back off, don't give them any room to breathe, or even look you in the eye. Crush them ruthlessly. The most important thing is to keep your enemy down; don't even give them the chance to get to their knees. Study this thoroughly.

21. Trading the Mountains for the Sea

In Japanese we pronounce 'mountains and sea' as 'sankai'. *Sankai* can also mean 'three times', so when I say 'trade the mountains for the sea', I mean that you should not repeat the same attack over and over again. Repeating the same move twice is natural, but never repeat it thrice. If you use a technique on your enemy and it is ineffectual, try it once again. If it is still not working for you, quickly change your

Samurai cavalry engage at the fourth battle of Kawanakajima in 1561.

Detail of part of a folding screen showing the siege of Osaka Castle (1615). The horned helmet indicates the figure of Honda Tadatome, one of Tokugawa Ieyasu's captains, as he leads an attack. Note the matchlock guns being aimed by Ashigaru *foot soldiers, which had been widely used in battle from the late 16th century.*

attack plan. If this next attack is unsuccessful, try another, different move. Storming your enemy like the sea when they are expecting mountains, and besieging them like the mountains when they are expecting the ocean is the Strategic Way that courses through this technique. Hone your skills in this well.

22. Knocking the Bottom Out

It may look at first glance as though you have applied the principles of the Way and won the fight, but the opponent's fighting spirit has not yet been broken. So although they look defeated on the surface, they have not yielded the fight in their heart. Knocking the bottom out addresses this. Quickly change your mindset, for it is crucial that you obliterate the enemy's spirit and show them defeat beyond what their mind can comprehend.

You can take their fighting spirit with your sword, with your body, or with your mind. There are many ways to achieve this. Once you have broken them, you can move on. But if they still have some fight left in them you must remain vigilant, for it is difficult to topple an enemy that still has spirit. Make sure you drill 'knocking the bottom out' fervently, for both large- and small-scale strategy.

23. Starting Over

This is necessary when you find yourself in a deadlock with your opponent. Throw away your current inclinations and begin anew in your mind, as though you are starting over. Feel the new rhythm and take your victory.

Starting over when you feel as though you are falling into a stalemate with your opponent means a rapid change of intent and use of an entirely different technique to win. This is also important to get into your head for a large battlefield, and with experience in the Way of the Warrior you will see it faster. Keep thinking on this.

24. A Rat's Head on an Ox's Body

When engaged with an enemy and you have both come to a standstill trying to think of all the particulars, remember that the Strategic Way of the Warrior is like the detail-oriented head of a rat on the audacious body of an ox. This is what 'a rat's head on an ox's body' refers to. It is the ability to suddenly change your mindset to be expansive when faced with detail and subtlety.

This switching between big and small is fundamental to the Strategic Way, so it is important for samurai to be able to employ this in everyday life. You should never depart from this way of thinking, as it is necessary in both large- and small-scale combat, so study it well.

25. The General Knows the Troops

'Knowing the troops' applies in any and all battles. With your continuous application of the Strategic Way and wisdom of strategy, you will be able to think of the enemy as your own troops to command. Think of manoeuvring them as you please, and they will be yours to direct freely. You are the general, and the enemies are your troops. Be mindful of this.

This 19th century woodblock print shows a samurai warrior in battle against monkeys.

A samurai celebrates victory with a ceremonial meal.

26. Letting Go of the Hilt

There are many ways to interpret this idea. It can mean winning without a sword. Or it can mean not winning with your *tachi*. It isn't possible to write down the numerous interpretations of this mindset, so train diligently to understand it.

27. A Body of Stone

The technique here, 'body of stone', comes with mastery of the Strategic Way. You will find that you are a being that is able to immediately embody the immense size and solidity of a boulder, untouchable and immovable. This will be taught in person.

Everything written above is just a vocalization of what has been in my thoughts constantly about my *Ichiryū* School of swordwork. It is the first time I am recording it all, so the order of articles might be somewhat muddled and the details expressed inelegantly. Nevertheless, it is all important for anyone wishing to walk the path and should be memorized.

Since I was a young boy I have devoted my life to the Strategic Way of the Warrior, sharpening my sword skills, honing my body through rigorous training and tempering my mind. I have explored the various other schools and their teachings, but have found them to be telling distorted stories, or caught up in the fine points of showy techniques, or even just performance arts. Not a single one possesses the true spirit of strategy.

Of course, it is possible to learn this kind of artistry, but practising and becoming accustomed to using your body and mind in this way will lead to bad habits in real strategy. These habits are hard to break and lead to the decay of the correct path. They will cause the Strategic Way of the Warrior to be slowly forgotten.

The true path of swordwork is knowing that winning a battle against your opponent and the principles of strategy are one and the same. If you take in all the wisdom of my Strategic Way and follow the true path, you will never need to doubt your victory.

By Shinmen Musashi-no-Kami-Genshin, 12 May 1645
To my disciple, Terao Magonojō

わが二天一流の兵法の知恵の力を体得し、正しく、真直ぐなありようを練習するならば、疑いなく勝ちを得るであろう

風の巻

The Scroll of Wind

Introduction

One must understand the Ways of other schools to know the Way of the Warrior. The Scroll of Wind will contain my record of their various strategies and trends. Without this understanding of other schools it will be truly difficult to grasp the components of my own *Ichiryū* School.

Looking at the other schools, there is a tendency to put great importance on things like wielding a bigger sword or swinging with as much force as possible. Some teach the Way through the use of a short style of *tachi* known as a *kodachi*. Yet other schools put emphasis on the many forms for sword practice, and some calling their stances 'extrinsic' and their use of the Way 'intrinsic'. In this scroll I will clarify how all of these schools are on the wrong path. I will reveal their shortcomings and strengths, and shed light on what their principles entail.

The principles of *Niten-Ichiryū* are something different entirely. These various other schools teach martial

This artwork shows samurai fighting up close using long katana *swords and* yari *spears.*

arts as a way of making a living, with florid displays of finesse that they somehow manufacture for sale. Do they think this is the right Way? The strategists of the world that view their training narrowly, purely as swordwork; merely practising the swing of a *tachi*, their agility, and their technical subtlety; do they think they have figured out the meaning of victory? Whatever they may think, none of this is the true Way. I will comb through the shortcomings of these other schools, one by one. Carefully reflect on each point, and use them for a full understanding of *Niten-Ichiryū*.

1. Swordwork in other Schools: Larger Blades

Among the other schools, there are those that favour a larger *tachi*. They should be considered weak schools from the perspective of my Strategic Way. This is because they have no comprehension of how to apply principles to win any fight, and instead rely on the advantage that sword length bestows when at a distance from the enemy. This is their only reason for preferring a longer blade. The well-known phrase 'size matters' is just an adage used by those who know nothing of strategy. Trying to take victory without going by the principles of the Way and instead relying on distance and a longer sword is only for the weak of heart. This is why I see it as a frail strategy.

When fighting in close range with an opponent, the longer the sword, the harder it is to strike with. If you can't wield it freely then it becomes dead weight. You will be even worse off than someone using a short *wakizashi* or just their

bare hands. Of course, to each their own for those who prefer the longer swords and like to listen to the adages. But when observed from the true path, such preferences will clearly not bring victory.

Will the shorter sword always lose the fight to the longer one? Depending on where you're fighting, in height-restricted or narrow places, or in settings where only the *wakizashi* would be permitted, reliance on a long *tachi* would directly contradict what the Strategic Way mandates. It is not a good mindset to prefer size over strategy, especially since there are those who are simply not strong enough to wield a long *tachi*.

I do not object to the longstanding phrase 'all shapes and sizes'. What I disagree with is someone bent on having their sword be longer. In large battles, the longer *tachi* is like having many troops, and the shorter *tachi* is like having few troops. Is it not perfectly possible for a smaller force to take on a larger one? The point of strategy is precisely that a smaller force can be victorious, and accordingly there are many examples dotted throughout history of a small force triumphing over a larger one. Bias towards a larger sword and a narrow mind is unacceptable in *Niten-Ichiryū*. Think this over carefully.

2. Swordwork in Other Schools: Stronger Blades

There is no such thing as a strong or weak *tachi*. A *tachi* swung with nothing but brute force makes for a rough cut, and it is impossible to win when one is rough around the edges. In fact, obsessing over a good strong blade and

*'The Loyal Ronin Crossing the Long Bridge to Embark for the Night Attack upon Moronao',
by Utagawa Hiroshige (circa 1840).*

Two samurai lock in close combat. The samurai who has the upper hand uses a wakizashi *short sword.*

swinging it with only your might will not cut anyone down. You wouldn't use such brute force when cutting through someone or something with a test swing.

No one, when locked in combat with an opponent, is thinking about whether to cut hard or cut softly. When trying to slay the enemy, it's not about a strong intent to cut, and certainly not about a weak intent to cut. Rather, it is about concentrating on your intent to kill. Moreover, thinking your blade is strong and smashing it too hard into the opponent's blade will only end badly for you. Hit their blade too hard and yours will shatter into pieces. This is why there is no such thing as a 'strong' blade.

In large-scale battle, if you have strong troops and the intent to achieve total victory, you must know that your enemy has also brought a strong force and intends to demolish you. Both sides are the same in this. That is why it is essential to have the logic and principles to win at anything. My school's path leads you to victory in any situation if you avoid obsessing over the inefficient and rely on wisdom from the Strategic Way. Study this in depth.

3. Swordwork in Other Schools: Shorter Blades

Trying to force victory using only short swords is not in line with the Way. The words *tachi* and *wakizashi* have existed to denote long and short swords since times past. There are strong people in the world who wield long swords with ease and have no need to harbour a preference for a short sword. In fact, they could readily wield even longer weapons like the spear or *naginata*.

With a shorter *tachi*, it is unwise to wait for your opponent to swing and then aim for an opening to cut them and grab their blade. Waiting for an opening will lose you the initiative and you will become entangled, which could lead to your downfall. Moreover, using a shorter sword and jumping into one enemy's space, when you are in the midst of multiple opponents, is not effective at all. Those disciplined in the short sword will swing about themselves when faced with multiple opponents, jumping about in a violent struggle. They end up parrying swords and getting swept up in the fray, but this is not the true Way. The most important thing for victory in this case is for you to remain strong and upright, while driving your enemy to jump about, flustering them into disarray. The principles for victory in large battles are the same. In this situation it is important to force enemy troops back and crush them on the spot.

Starting to learn the Strategic Way by learning how to block, dodge, disengage and deflect attacks from the outset ingrains a defensive mindset and gives the enemy the control to manipulate you. The Way of the Warrior, on the other hand, is straight and true, and with the right principles allows you to drive the enemy to subjugation. This is important to remember.

4. Swordwork in Other Schools: Multitudes of Sword Techniques

Peddling a plethora of forms and techniques exploits the Way as a profitable commodity, and teachers purposely use a multitude of techniques to make students buy into how profound the training is. This is deplorable. The reason it is in opposition with the Way is because it only confuses the

This hand-coloured illustration from a Japanese miscellany on traditional trades, crafts and customs in mid-18th century Japan shows a naginata *in the foreground. The* naginata *was used by both samurai and* ashigaru *foot soldiers.*

mind to think that there are many ways to cut someone. The ways to cut someone that exist in the world are not in flux. Regardless of whether the wielder of the sword is a master, a beginner, a woman or a child, the number of ways to strike and cut a person does not change. Even if we depart from cutting as a technique, there is nothing else except stabbing and slashing. There can't be hundreds of techniques if the single purpose is just to cut someone.

Having said that, depending on the battleground and the circumstances, overhead or out to the side might be restricted, making the *tachi* difficult to use. To this end, the five ways to hold a sword exist to deal with that. Other than those, however, cutting by contorting your wrists or body, or by leaping around is not concurrent with the correct Way. Twisting won't cut, nor will swivelling, leaping or opening the body position. All of these are completely futile.

In *Niten-Ichiryū*, remember it is important that your own body and mind should remain straight while you bend the enemy around and win while their mind is still reeling. Think this over.

5. Swordwork in Other Schools: Use of Stances

It is a mistake to put the sword stances above all else. Stances exist because there is no enemy to fight right now. Put another way, making up new precepts based on the customs of old or the current system should not happen for matches held in accordance with the Strategic Way. Stances should be used to work the enemy into a compromised position.

A battle takes place between the villainous samurai Akabori Mizuemon and the female warrior Omatsu.

139

Taking up a stance means putting your resolve into an unshakeable posture. It is like positioning a castle or an army to be impregnable, even when besieged by enemy forces. However, this is still for times of peace. In a contest in the Strategic Way, you should constantly strive to get that initiative, but by taking a stance you are stood waiting for the opponent to take the lead. Work this out through study.

The skills we employ in the Strategic Way involve shifting the opponent's stance, attacking in unexpected ways, bewildering them, shaking up their movements, intimidating them, and using the interrupted rhythms of their unsettled mind to win. Therefore, giving up the initiative by adopting a stance is unacceptable. For this reason, the stances in *Niten-Ichiryū* are 'stanceless stances' and are defined yet flexible positions.

When fighting in large battles, from the outset it is important to know the enemy's numbers, take in the conditions of the battlefield, understand the way your troops work best and make use of their strengths when placing them.

Taking the initiative and attacking first is twice as effective as waiting for the enemy to come to you. Assuming a defensive stance and preparing to parry or deflect the oncoming attack is like making a fence out of spears and *naginatas* and letting the enemy equip themselves. Instead, pull those fence posts out and use the spears and *naginatas* as devices against the enemy. Think about all of this thoroughly.

Three Ronin samurai attacking the entrance to Morono's (Kira Yoshinaka) home. From Juichidanme, Act Eleven of the Chushingura, *a woodcut print by 19th century artist Utagawa Kuniyasu.*

Samurai warriors on the march, from a 19th century illustration.

6. Swordwork in Other Schools: Fixing the Gaze on a Point

The gaze, according to some schools, should be fixed on the enemy's *tachi*. Other schools teach students to spot the hands or the face, or even the feet. However, doing so will result in an overly fixated mindset and problems in your strategy.

For example, people playing keepie-uppie (kick-ups) don't keep their eye fixed on the ball. A well-practised player won't be looking at the ball as they kick it after a rebound from the temple, or while doing high-level tricks like a sole kick or a spin kick. This goes for jugglers as well. They practise so much that they don't need to doggedly keep staring at the door balanced on their nose, or the many swords they juggle. They just learn to look as needed in the course of their practice.

Similarly, when you grow accustomed to fighting many enemies in the Strategic Way, you will be able to see if the enemy is cautious or careless; if they raise their swords at you, you will see if they're coming in to close the distance, or keep it, and how fast. If you keep your gaze fixed anywhere, it is on the enemy's intent or, in large-scale combat, on the size and status of the enemy force.

The ways to 'look' at the detail near you and 'see' the breadth around you are related to this. Intensify the way you see into the enemy's mind while also looking at the state of the area around you. Keep the overall progress of the fight in your vision and look out for the strengths and

weaknesses all over. This is, without a doubt, important for claiming victory.

There is never a time when someone should keep their gaze fixed on the trivialities. I have written this previously, but if you only have eyes for the minute details you will forget the big picture and become uncertain, and your victory will completely escape you. Think about this principle deeply and continually build experience in it.

7. Swordwork in Other Schools: Multitudes of Footwork Techniques

There are various ways to move the feet swiftly according to other schools. Among the ways to move are floating feet, jumping feet, prancing feet, heavy feet, staccato feet and crow's feet. All of these are insufficient when attempted in line with *Niten-Ichiryū*.

I reject floating feet. The reason for this is that a warrior will habitually float their feet in battle, when this is a path that requires your feet to be firmly on the ground, whatever happens. Jumping feet is not preferable because the action of jumping itself takes power and leaves you open for the fraction of a second that your feet adjust to landing. There is no reason to jump up and down, so jumping feet as a technique is useless to you. Prancing feet will also get you nowhere while you concentrate on the prancing itself. I am particularly against the use of staccato feet because it causes you to stop in place and wait.

Other than that, there are many swift-footed techniques like crow's feet. However, think about being in a marsh,

Matsumoto Kōshirō III, one of the most famous 18th century kabuki *actors, adopts a low posture, sword at the ready.*

or deep rice fields, mountains and rivers, rocky plains or narrow roads. There are so many environments in which you may have to face your opponents and you won't be able to jump or prance, or move at speed over all of them.

There is no change to the movement of your feet in the Strategic Way. It is enough to walk as you always walk in the street. Adjust your posture to counter the haste of the enemy's rhythm with a calm bearing, ensuring that you don't give too much or too little so that your steps don't falter.

Your footwork is very important in large-scale battle. What I mean by this is that victory will not come easily if you rush in gung-ho without due care to what the enemy is intending, as your rhythms will be out of time. Equally, if you drag your feet in making your offence, you won't find where the enemy is faltering at breaking point and you'll lose the opportunity to strike fast and finish the contest. It is vital to use the moment when the enemy is crumbling to win without giving them even a moment to breathe. Build up your training over time.

8. Swordwork in Other Schools: The Use of Speed

Speed is not pertinent to the Strategic Way. Uncertainty stems from not being well aligned with the rhythm of things, and this is what creates speed or sloth. A person talented in their path will not appear hasty.

For example, there are messengers that cover 160 to 200 kilometres (100 to 125 miles) in a day, but they don't run at full tilt from morning till night. Inexperienced messengers

A samurai moves at speed, his armour peppered with his enemy's arrows.

A samurai should be prepared and not act in haste.

might run at speed all day but will make no progress. A *Nōh* theatre novice singing with an expert will get flustered trying to keep up. The *taiko* drum song 'Old Pine' is a relaxed song, yet a beginner may feel as if they're lagging behind and end up speeding through. The tempo of the song 'High Sand' is fast, but to play it hastily is wrong. The notes end up tumbling out in the wrong time. But of course, playing it too slowly is also incorrect. A skilled player appears relaxed while not missing a beat. Whatever the case, a master of their craft never shows haste in their movements.

Take from these examples an understanding of some principles in the Strategic Way. Speed is bad in strategy. That is, depending on the place, a marsh or a rice field will make quickly moving your body or feet nearly impossible. Going for a swift strike with your *tachi* is also bad. By attempting to cut something with speed, instead of the sword moving like a fan or dagger, it will actually not cut anything at all. Consider this carefully, as speed and haste are also unnecessary in large-scale battle.

Stick to 'nipping it in the bud' and you will never be behind. Furthermore, when the enemy is coming at you with senseless speed, remember to retaliate with a sense of calm and do not be drawn into their haste. Think this over and train hard to understand the attitude for this.

9. Swordwork in Other Schools: Inner and Outer Teachings

In the Strategic Way, what is on the surface and what is underneath? Depending on the art, there are inner and

山奥からさらに奥へ行こうとす

outer 'secret teachings', but when you are fighting an opponent there is no such thing as fighting with surface techniques and cutting with inner teachings.

When I teach my Strategic Way of the Warrior to new students, I start them learning the techniques that are easy to apply. Their mind will naturally open to gradually more advanced techniques, and as I see that happening I begin to teach deeper principles. However, I generally change my approach based on the condition of the learner, so there is no distinction between inner and outer teachings here.

We say that, 'When venturing deep into the mountains, trying to go ever deeper will lead you back around to the entrance.' Whatever the path, sometimes the deeper matters are more pertinent, and sometimes the surface matters are better. In this particular theory of combat, how can we say what to hide, and what to show on the surface? That is why when I teach my Way, I do not require some written oath swearing students to secrecy with the threat of punishment, should they divulge anything. I simply weigh up the student's ability to study the path, then teach them the correct Way, and have them throw out their bad habits gained from their time in the Six Realms of Strategy, which are as dangerous as the Six Buddhist Realms of Existence.

My Strategic Way is taught by guiding the student to naturally enter the correct path for a warrior and by freeing their mind of all doubt. This is how it should be, and you must take the time to train yourself.

This 18th century woodblock print depicts actor Segawa Kichiji II as a daimyō's young son being served by an attendant.

An actor plays a young samurai in this 18th century woodblock print by Torii Kiyotada.

Above is the culmination of what I have chosen to record in the Scroll of Wind. I have split the strategies of the other various schools into nine sections. Although there are many facets of the inner and outer teachings of the schools I could have made clear, I have decided to leave out the names of the schools and their individual priorities.

The reason for this is that each school has their own way of thinking and different explanations, as does each person have their own understanding. Even within the same school, the interpretations may be subtly different. To ensure the record does not become antiquated if things change, I did not identify individual schools, and instead split the main ideas into the nine sections.

Looking at them from a worldly, or objective person's viewpoint, these schools all have their own biases; whether it's for a longer or a shorter sword, for strength or weakness, for a breadth of techniques or a focus on detail. Because each has their own tilt on the path, even without me defining the ins and outs of each school, everyone should know what they are.

In *Niten-Ichiryū* there is no distinction between inner and outer in swordwork. There are no set stances. Simply working your mind with integrity and perceiving the virtues of strategy is the substance of the Way of the Warrior.

By Shinmen Musashi-no-Kami-Genshin, 12 May 1645
To my disciple, Terao Magonojō

空の巻

The Scroll of Expanse

I record the *Nitō-Ichiryū* Strategic Way of the Warrior within this Scroll of Expanse.

The spirit of Expanse is a place that contains nothing, one that is infinite and incomprehensible, much like the sky itself. Of course, the Expanse itself is also nothingness. Understanding the existent gives an understanding of the non-existent. That is the Expanse.

The world mistakenly views Expanse as the inability to comprehend certain things, but that is not the truth of the Expanse. This view is nothing but an uncertain heart. Nor is it the Expanse to walk the path of the Strategic Way of the Warrior as a samurai, yet fail to understand the warrior's code. Having all manner of uncertainties and calling that which you cannot understand 'Expanse' directly conflicts with what Expanse truly means.

True samurai must memorize the Strategic Way of the Warrior and give earnest effort in other arts, while not becoming overshadowed in their practice of the Way.

They must keep an unwavering spirit and be vigilant in every hour of every day. They must polish the mindset of perception and the mindset of concentration, sharpen their vision for looking and seeing, and keep an unclouded view. Only in the clarity brought by clearing away the clouds of doubt can the warrior understand the Expanse.

People stray from their paths if they are ignorant of the truth, be it the path of Buddhism or another worldly path, even if they believe they are doing the right thing. It is clear from the perspective of an unfettered spirit and the universal principles of the world that they have strayed because of the favouritisms in their heart and a tilted view of things.

Understand this notion and root yourself in uncompromised truths, practise the Way extensively, with authenticity and clarity, with the true spirit as your guide. Make the Expanse your Way and you will see the Way is your Expanse.

The Expanse is of virtue, not vice.
Wisdom is of substance.
Logic is of substance.
The Way is of substance.
Your spirit is the Expanse.

By Shinmen Musashi-no-Kami-Genshin, 12 May 1645
To my disciple, Terao Magonojō

Glossary

Artisan
A worker in a skilled trade, especially one that involves making things by hand.

Buddhism
Buddhism is a religion that was founded by Siddhartha Gautama ('the Buddha') more than 2500 years ago in India. Followers of Buddhism don't acknowledge a supreme god or deity. They instead focus on achieving enlightenment – a state of inner peace and wisdom. When followers reach this spiritual echelon, they are said to have experienced nirvana.

Bunbu-Ryōdō
A traditional motto of the Japanese Samurai who cultivated their personal development in addition to their martial arts training and expertise with the sword. *Bun* means 'culture' or 'literacy', while the second character *bu* is usually translated as 'fighting' or 'martial power'. In *Ryodo*, the first character *ryo* means 'both', while the last character *do* means 'way'. The component characters are merged together to convey the overall meaning of 'cultural and martial spirit or prowess'.

Confucianism
Confucianism is a way of life taught by Confucius (Kong Fuzi) in China in the sixth – fifth century BCE and the rituals and traditions associated with him. Sometimes viewed as a philosophy, sometimes as a religion, Confucianism is perhaps best understood as an all-encompassing humanism that is compatible with other forms of religion.

Disciple
A person who believes in the ideas and principles of someone famous and tries to live the way that person does or did.

Duel
In a culture pervaded by honour, perceived slights against that honour can be remedied with a duel to restore the balance or decide who truly has the higher honour through their actions.

Katana
A curved sword with a blade length of more than 0.6m (2ft). It varied slightly from a *tachi* sword. It was a marker of samurai status in the 16th century.

Kodachi
A *kodachi* literally translates into 'small or short *tachi*'; this Japanese sword was too short to be considered a long sword but too long to be a dagger. Because of its size, it could be drawn and swung extremely fast.

Kyoto
Kyoto is the capital city of Kyoto Prefecture in Japan. Located in the Kansai region on the island of Honshu, it served as Japan's capital and the emperor's residence from 794 until 1868.

Lancers
Those that use spears.

Musketeer
A soldier armed with a musket.

Naginata
A weapon with a long curved blade attached to a wooden staff.

Nōh
A form of traditional Japanese dance-drama theatre performed by actors in masks. It is highly stylized with costumes, sets, props, masks and characters that are based on traditions.

Sake
An alcoholic drink made from fermented rice. The foundations of good sake are quality rice, clean water, koji mold and yeast. They are combined and fermented in precise processes that have been refined over the centuries.

Samurai
Originally the dependent followers of *gokenin* (warriors who performed guard duty for the Kamakura regime), this became a general term for all warriors in the 17th century.

Shintō-Ryū
A classical style of swordsmanship that was founded in 1447 by a man named Iizasa Chosai Ienao. It consists of a number of different skills such as sword, staff, halberd, spear and the combined use of the long and short sword.

Tachi
A sword. During the 14th century long-words and, to a lesser extent, *naginata* were the preferred weapons for hand-to-hand combat. But, the bow remained the dominant weapon throughout the 14th century. Varied slightly from the more curved blade of a *katana*.

Taiko
A drum made from an open-ended wooden barrel with animal skin stretched over both ends. The drum is played with two *bachi* (wooden sticks).

Tajima
A region located in the northern part of Hyogo prefecture. It includes the cities of Asago, Yabu and Toyooka and the county of Mikata. The northern part faces the Sea of Japan.

Tanegashima
A type of matchlock configured musket firearm introduced to Japan through the Portuguese in 1543 *Tanegashima* were used by the Samurai and their foot soldiers and within a few years the introduction of the *Tanegashima* in battle changed the way war was fought in Japan forever.

Yin **and** *Yang*
The principle of *Yin* and *Yang* is that all things exist as inseparable and contradictory opposites, for example, female-male, dark-light and old-young. The principle, dating from the third century BCE or even earlier, is a fundamental concept in Chinese philosophy and culture in general.

Waka
A type of poetry in classical Japanese literature.

Wakizashi
A short curved sword with a blade generally from 0.3–0.6m (1–2ft) in length.

Index

adhering blades 72
antagonizing 110
autumn leaves strike 66, 69

battles and contests
  acting in infectious ways 109
  antagonizing 110
  body of stone 125
  chipping away at the corners 112, 115
  crushing the enemy 118
  detecting the collapse 103
  flustering the enemy 115
  fusing with the enemy 112
  high ground 88
  interweaving 117–18
  invoking fear 110, 112
  knocking the bottom out 121
  knowing the conditions 99
  knowing the troops 122
  letting go of the hilt 125
  lifting the shadow 107
  nipping it in the bud 95, 97
  releasing the four-handed grip 104, 107
  shouldering the sun 86, 88
  starting over 121–2
  stifling the shadow 107, 109
  the three cries 115, 117
  the three initiatives 91–2, 95
  trading the mountains for the sea 118, 121
  traversing pivotal points 97, 99
body of stone 125
body over sword 69
body slamming 72

crushing the enemy 118

double-beat rhythm 63, 65

embodiment of lacquer and glue 71
embodiment of the autumn monkey 71

fear, invoking 112
Five Forms, the 57, 59–60, 62
Five Scrolls, definition of 22, 24, 27–8
Five Stances, the 54, 59–60, 62
flint and steel strikes 66
flowing water strikes 65
flustering the enemy 115
footwork 52, 54, 65, 144, 146
four-handed grip, releasing 104, 107
four paths of the samurai 16, 18
fundamental rules 41, 43
fusing with the enemy 112

gaze 51, 143–4

high ground 88

inner and outer teachings 149–50, 153
interweaving 117–18

knocking the bottom out 121
knowing the troops 122

'lash-slash' stab and cut 76
letting go of the hilt 125
lifting the shadow 107

measure of height 71
mindset 47–8
multiple opponents, engaging 77, 79

nipping it in the bud 95, 97

occupations 16, 18
opportunity strikes 66

parries 72, 75, 76–7
posture *see* stance
principles of engagement 79

rat's head on an ox's body 122
Rhythm of the Way 38, 41
rules, the 41, 43

schools of the Samurai 27, 28, 129–153
shouldering the sun 86, 88
'single beat strikes' 63
single blows 79
speed, the use of 146, 149
stabbing the chest 75–6
stabbing the face 75
stance 48, 51
  body slamming 72
  embodiment of lacquer and glue 71
  embodiment of the autumn monkey 71
  Five Stances, the 54, 57, 59–60, 62

measure of height 71
in other schools 138, 140
stanceless stance 62–3
starting over 121–2
stifling the shadow 107, 109
Strategic Way of the Warrior 11–28
strikes
  adhering blades 72
  autumn leaves strike 66, 69
  body over sword 69
  engaging multiple opponents 77, 79
  flint and steel strikes 66
  flowing water strikes 65
  'lash-slash' stab and cut 76
  multitudes of sword techniques 136, 138
  opportunity strikes 66
  parries 72, 75, 76–7
  single-beat strikes 63
  single blow 79
  stabbing the chest 75–6
  stabbing the face 75
  the 'strike' and the 'hit' 69
success, importance of 12, 15
swords, samurai (*tachi*)
  double-beat rhythm 63, 65
  Five Forms, the 57, 59–60, 62
  footwork 52, 54
  the grip 52
  larger blades 131–2
  multitudes of sword techniques 136, 138
  principles of engagement 79
  shorter blades 135–6
  stronger blades 132, 135
  swing 54, 57
  two swords as one 28, 31, 33
  see also stance; strikes

tachi see swords, samurai (*tachi*)
Ten Skills and Seven Arts, the 15
three cries, the 115, 117
three initiatives, the 91–2, 95
trading the mountains for the sea 118, 121
training, consistency of 41, 43
traversing pivotal points 97, 99
True Path, the 54, 57
'tying the linchpin' 51

unconscious, unpredictable strikes 65
weapon variations, advantages of 34, 37–8
wisdom enriching 48

Picture Credits

Alamy: 17 (Old Books Images), 30 (Interfoto), 40 (James Tangkilisan), 42 (Heritage Image Partnership), 49 (Artefact), 61 (CPA Media), 70 (BTEU/RKMLGE), 94 (CPA Media), 96 (Mark Edward Eite/Aflo), 98 & 101 (World History Archive), 102 (The Protected Art Archive), 105 (The Picture Art Collection), 106 (Science History Images), 108 (The Protected Art Archive), 111 (CPA Media), 114 (Science History Images), 123 (Photo 12/Ann Ronan Picture Library), 124 (Chronicle), 130 & 137 (CPA Media), 141 (Artokoloro), 142 (Historica Graphica Collection/Heritage Images), 147 & 148 (The Protected Art Archive)

Amber Books: 9, 27, 45, 75, 83, 85, 97, 117, 127, 129, 155

Getty Images: 25 (USC Pacific Asia Museum), 32 (DEA/A Dagli Orti), 58 (Historica Graphica Collection/Heritage Images), 68 (Art Media/Print Collector), 78 (Heritage Images), 82 (Art Media/Print Collector), 93 (Sepia Times/Universal Images Group), 116 (Archiv Gerstenberg/ullstein bild), 120 (Werner Forman/Universal Images Group)

Library of Congress: 5, 7, 13, 29, 35, 36, 46, 56, 64, 67, 73, 113, 119, 134

Metropolitan Museum of Art, New York: 10, 14, 20, 26, 39, 50, 53, 55, 74, 81, 87, 133, 139, 145, 151, 152

Public Domain: 23, 90

Shutterstock: 89 (djgis)